Theory and Applications of Mechanical Sciences (Part-2)

CONTENTS

SECTION A : BASIC THERMODYNAMICS

DEDICATED TO

LOVING MEMORY OF LATE GUNAMOY ADHWARJEE.

DIRECTOR
☎ : Office : (0343) 2546397
Fax : (0343) 2547375/2546753
E-mail : director@nitdgp.ac.in
dgp_recdgppr@sancharnet.in
Website : www.nitdgp.ac.in

राष्ट्रीय प्रौद्योगिकी संस्थान दुर्गापुर
NATIONAL INSTITUTE OF TECHNOLOGY DURGAPUR
DEEMED UNIVERSITY
MAHATMA GANDHI AVENUE, DURGAPUR - 713209 (WEST BENGAL)

Dr. M.C. Majumder
Professor & H. O. D.
Mechanical Engg. Dept.
N. I. T. Durgapur
Durgapur 713 209

Date: 11. 11.07

Prof. Dilip Kumar Adhwarjee
Dr. B.C. Roy Engineering College
Durgapur - 713 206

FOREWORD

The book "Theory and Applications of Mechanical Sciences (Part 2)" written by Prof. Dilip Kumar Adhwarjee, is a good addition to the existing list of books on similar topics. With his long years of industrial and academic experience, Prof. Dilip Kumar Adhwarjee has produced concepts in a more lucid manner. He has used his class notes during of his study at REC Durgapur (NIT) and updated those for the students of the West Bengal University of Technology (WBUT). I wish the concise treatment of the topics of Mechanical Sciences (Part 2) will prove beneficial to the under graduate engineering students because of its strict adoption of WBUT syllabus.

With best wishes

11/11/07

Dr. M. C. Majumder

PREFACE

I am delighted to present the book entitled. "Theory and Applications of Mechanical Science (Part 2)" to the engineering students of all branches. I have taken special care to present the subject matter in a simple manner.

Some salient features of this book are :

- A large number of numerical problems are completely solved.
- The contents have been organised in a systematic order.
- This book can be used as self-study.
- The languages is simple and easy to understand.
- Figures and diagrams are neatly drawn so that students can understand text easily.

Hence every possible attempt has been made to make this book useful for the engineering students. All the students will be able to grasp the subject by self study. Any constructive suggestions and criticisms regarding the improvement of this book will be appreciated.

—Author

ACKNOWLEDGEMENT

I express my gratitude to the professors of N.I.T. (R.E.C), Durgapur who taught me Mechanical Engg. Subjects during 1975–1980. I have used their class notes in my book.

I am thankful to all the teachers and ex-teachers of Maliara R.N. High school, who taught me Physics and Mathematics which are the basis of this book.

Finally, I acknowledge the best support and encouragement from Shri Dulal Mitra, Secretary of Dr. B.C. Roy Engineering College, Durgapur.

—Author

SECTION A

BASIC THERMODYNAMICS

Chapter 1

Introduction

1.1 INTRODUCTION TO THERMODYNAMICS

Thermodynamics is the thermal science dealing with energy transfer in the form of heat and work and their conversion to each other. There is always some effect of energy transfer on the physical properties of the substance. Human desires output as work and he utilizes heat as an input. Thermodynamics tells us the most economical way of utilizing heat to produce work and vice-versa. The thermal science was originated in 19th century when steam engine was introduced. The size of this engine was large and the efficiency & performance were low. For better performance and efficiency, steam turbines, I.C. engines, gas turbines were invented in last few decades.

1.2 MACROSCOPIC AND MICROSCOPIC POINT OF VIEW

Fig. 1.1(*a*) shows a cylinder having air and fuel at definite volume, pressure and temperature. The piston is now pushed to level 2-2, and at the new situation there is definite new volume, pressure and temperature.

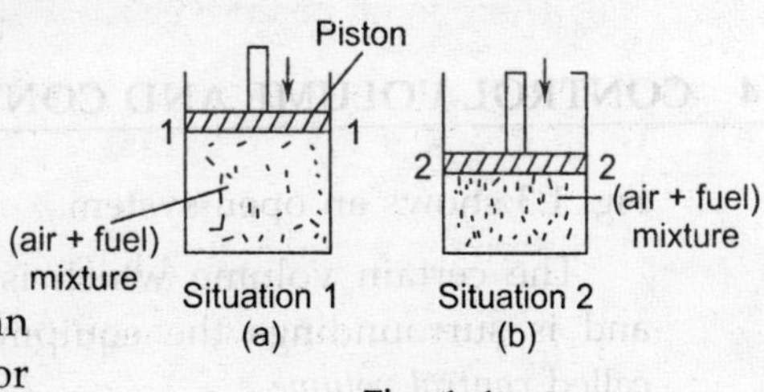

Fig. 1.1

The volume, pressure and temperature can be measured in both the situations. The state or condition of the systems (Air + Fuel mixture) can be determined by means of above characteristics or properties. This type of analysis is macroscopic analysis and is used in engineering thermodynamics.

However, if the analysis is made on the basis of the atomic or molecular behaviours of the substance, the analysis is called microscopic analysis. It is to be noted that macroscopic results can also be derived from microscopic analysis.

1.3 CONCEPTS OF SYSTEMS

A thermodynamic system (Fig. 1.2(*a*)) is defined as a fixed quantity of matter or a region in space upon which attention is focussed for the study of thermodynamic properties of the matter or substance. Everything outside the system, which interacts with the system is called surroundings or environment. The interface between the system and surrounding is the boundary. The boundary may be real or imaginary and may be fixed or moving. Mass and energy transfer can take place through boundary.

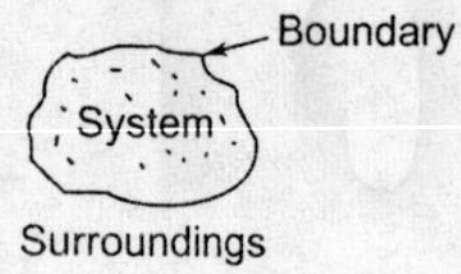

Fig. 1.2 (a)

There are three types of system.

- Closed System
- Open System
- Isolated System.

1.3.1 Closed System

A system in which the mass is constant and only energy can flow through the boundary in or out, is called a *closed system*.

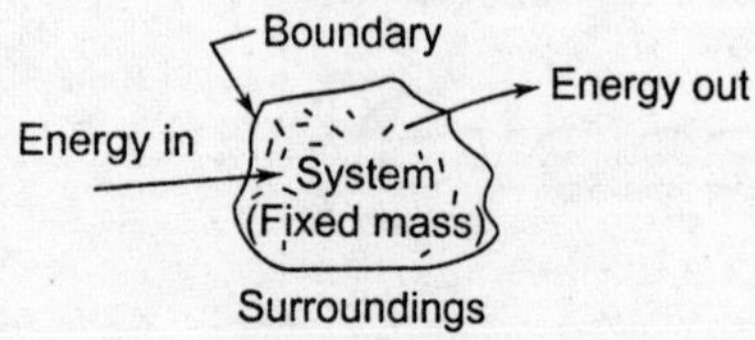

Fig. 1.2 (b)

Fig. 1.2 (*b*) is the closed system. Here there is no mass transfer.

1.3.2 Open System

A system in which the mass crosses the boundary in addition to heat and work transfer is called an *open system*. Fig. 1.2 (*c*) is an open system.

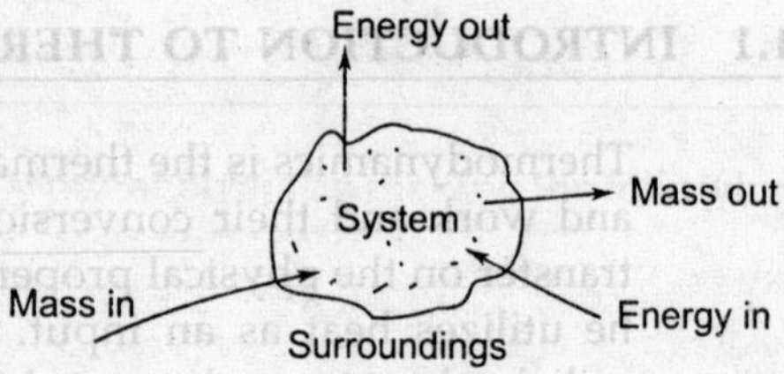

Fig. 1.2 (c)

1.3.3 Isolated System

A system where there is no interaction between the system and surroundings is called *isolated system.* There is no mass transfer and no energy transfer. Therefore it is the fixed mass and fixed energy system. Fig. 1.2 (*d*) is an isolated system.

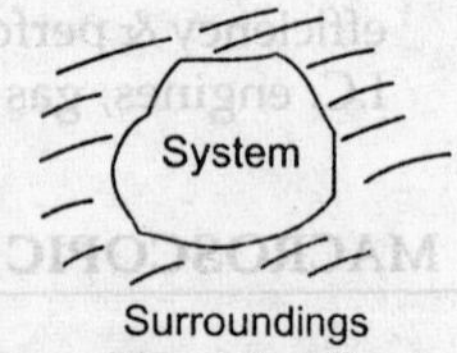

Fig. 1.2 (d)

1.4 CONTROL VOLUME AND CONTROL SURFACE

Fig. 1.3 shows an open system.

The certain volume which is fixed and is surroundings the equipment is called *control volume*.

The fixed boundary enclosing control volume is called *control surface*.

Regardless of the type of system, the control surface must be specified for the solution of any problem.

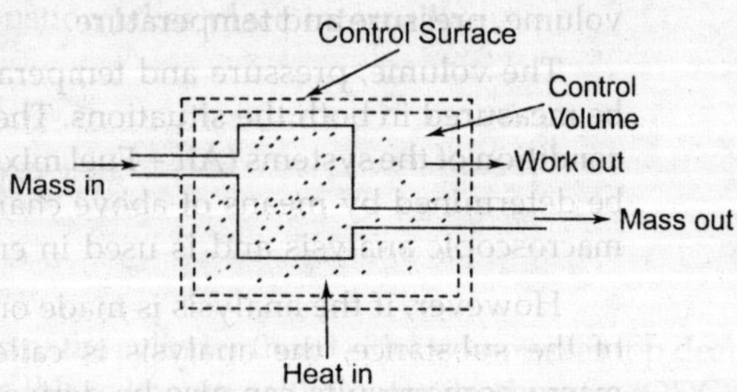

Fig. 1.3

1.5 HOMOGENEOUS AND HETEROGENEOUS SYSTEMS

A system is said to be *homogeneous* when the matter within the system exists in a single phase with constant chemical composition and physical properties. For example, water, air etc.

A system which consists of more than one phase is known as *heterogeneous* system. Air + fuel mixture in a Petrol engine cylinder is *heterogeneous* system.

1.6 PROPERTY AND STATE

1.6.1 Property

The characteristics of a substance which are measurable or calculated is called property of the system. For example; pressure, volume and temperature which are macroscopic in nature. The properties which are independent of mass, such as pressure, temperature, specific volume are called *intensive* properties.

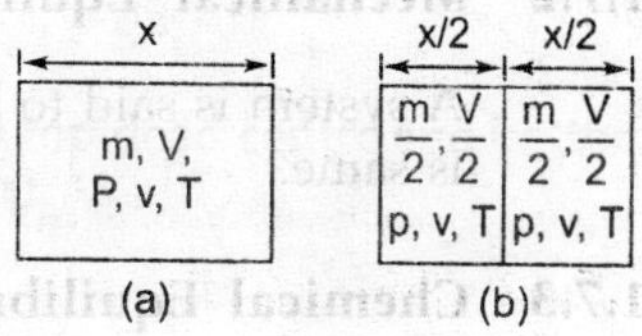

Fig. 1.4

The properties which are dependent of mass such as volume, energy are called extensive properties.

Fig. 1.4. (*a*) shows a mass of substance (*m*), volume V; and temperature T. A diaphragm is put in the mid way of the container. The two chambers (Fig. 1.4 (*b*)) thus formed have volume $\frac{V}{2}$, mass $\frac{m}{2}$ but P, T and specific volume v is the same. So V denotes extensive property and P, T and v denote intensive properties of the substance.

1.6.2 State

A state is the condition of a system which has definite properties and when can be represented at a point in the thermodynamic property coordinates such as (P, v) coordinate.

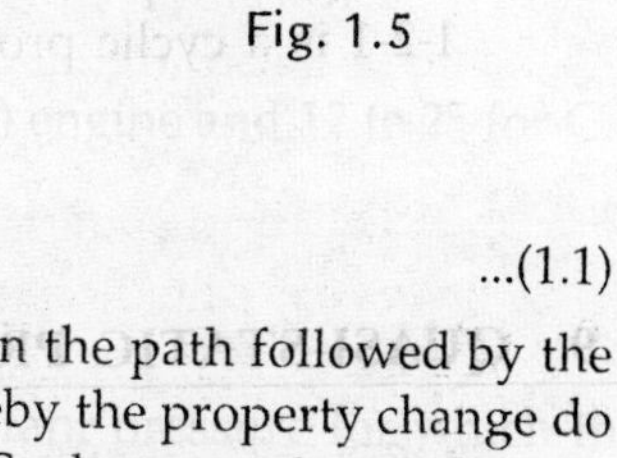

Fig. 1.5

Fig. 1.5 shows point 1 which is at state (P_1 v_1) and 2 at (P_2 v_2). When the system undergoes a change from state 1 to state 2, heat or work transfer will be there. The change in the absolute value of the property between the state 1 and state 2 is obtained from the difference between the coordinates of the two points representing the states.

$$\therefore \quad \Delta P = P_2 - P_1, \text{ and } \Delta v = v_2 - v_1 \qquad ...(1.1)$$

Equation (1.1) shows that the properties do not depend on the path followed by the system during the change but depend only on the state. Thereby the property change do not depend on the paths A, B, C but on end points (1 and 2). Such properties are known as point functions. Any variable is a property if it has a definite value at each state of equilibrium.

1.7 THERMODYNAMIC EQUILIBRIUM

A system is said to be in thermodynamic equilibrium when there is no change of properties of the system if isolated from surroundings. There are three types of thermodynamic equilibrium.

- Thermal Equilibrium
- Mechanical Equilibrium
- Chemical Equilibrium.

1.7.1 Thermal Equilibrium

A system is said to be in thermal equilibrium when the temperature at all points in the system is same.

1.7.2 Mechanical Equilibrium

A system is said to be in mechanical equilibrium if pressure at all points in the system is same.

1.7.3 Chemical Equilibrium

A system is said to be in chemical equilibrium if there is no chemical reaction, diffusion or any transfer of matter.

1.8 PROCESSES AND CYCLES

1.8.1 Process

When a system undergoes a change of state, a *process* occurs. The succession of states passed through during a change of state is called path. The path is completely specified in a process. For example, a constant volume process, a constant temperature process etc.

1.8.2 Cycle

A thermodynamic *cycle* is defined as a number of processes in sequence where the final state is identical to its initial state.

Fig. 1.6 shows A – B as a process and 1-2-1 as a cycle. To complete a cycle, there may be number of processes in between. When the system comes to its original state after passing through a sequence of processes, then it is called a *cyclic process.* 1-2-1 is a cyclic process.

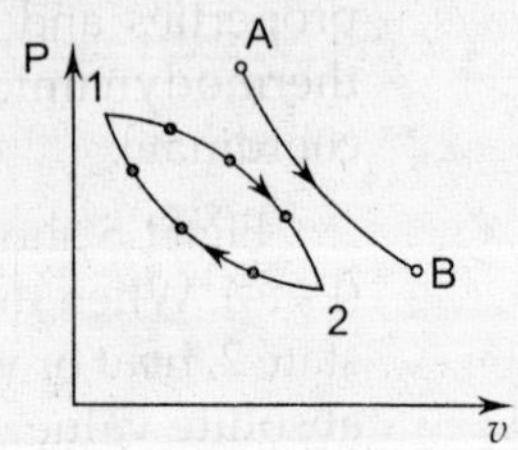

Fig. 1.6

$$\oint dP = 0, \oint dv = 0$$

1.9 QUASI-STATIC PROCESS

Properties are described completely when the states of a system is in thermodynamic equilibrium. *How can we describe the intermediate states of a system along the path of a process since sthe process takes place only when the system is in equilibrium?* We can justify the above question by the following example. Let us take a piston-cylinder mechanism as shown in Fig. 1.7. The weight W is made up of several small weights like W_1, W_2, ..., W_9. If we remove the weights one by one slowly, then the gas inside the cylinder undergoes infinitesimal change of state. All the states in between will be deviated from equilibrium infinitesimal. This intermediate processes are called *quasi-static process* or *quasi-equilibrium process.*

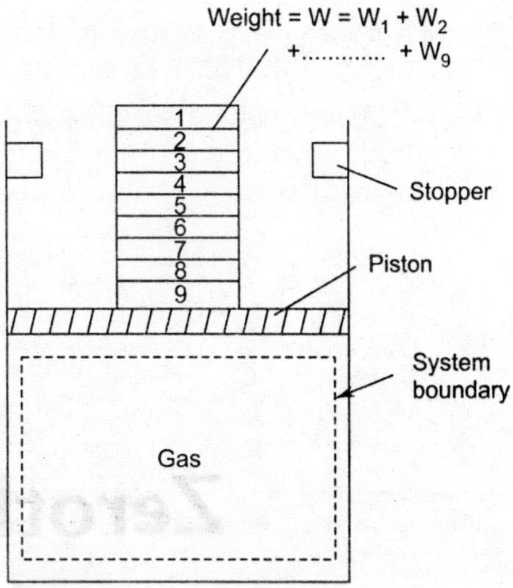

Fig. 1.7

However, if the total weight W is removed suddenly, the piston will get a jerk and strike the stoppers. It will come to equilibrium after many oscillations. But in this case, it is not possible to describe the intermediate states or paths. Such a process is a non-equilibrium process.

Thus in quasi-static process, there is only an infinitesimal change in properties of substance from its equilibrium.

In several cases, one property remains constant during a process. The prefix iso is used to describe such a process. If temperature remains constant, the process is called *isothermal process.* If pressure remains constant, the process is known as *isobaric* or *isopiestic process* and if volume is constant, it is called *isochoric process.*

1.10 REVERSIBLE AND IRREVERSIBLE PROCESSES

1.10.1 Reversible Process

If a process can be completely reversed, it is called *reversible process.* Fig. 1.8 shows a process 1 – 2 in the forward direction, if the process is carried out in the opposite direction (backward direction) *i.e.,* from 2 to 1 in the same succession of states as followed in the forward direction, it is called a reversible process. Thus, the system comes to initial state. In addition, the interaction between the system and surroundings are also identical but opposite in direction. Thus the surrounding also comes to its initial state. A quasi-static process is a reversible process.

1.10.2 Irreversible Process

A process which is not reversible is *irreversible process.* A quasi-static process with friction is irreversible process. Non-equilibrium process cannot be exactly reversed and the system cannot be brought back to the original state without leaving a net change in the surroundings.

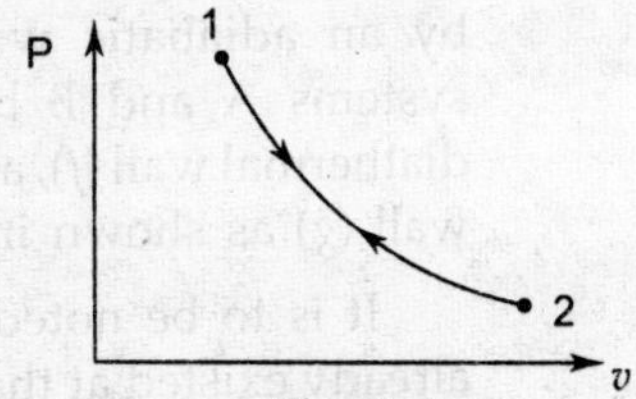

Fig. 1.8

Chapter **2**

Zeroth Law of Thermodynamics and Temperature

2.1 STATEMENT OF ZEROTH LAW OF THERMODYNAMICS WITH EXPLANATION

"When two systems A and B have thermal equilibrium (equality of temperature) with a third system C, then they (A and B) in turn have thermal equilibrium (equality of temperature) with each other".

To explain the above statement let us consider three systems A, B and C as shown in Fig. 2.1 (*a*), enclosed in a rigid adiabatic wall (*r*). Let systems A, B and C coupled to system C via diathermal wall (*e*). A is separated from B by an adiabatic wall (*d*). Then it is natural that A and B will be separately in thermal equilibrium with C. Practically it is seen that A will be in thermal equilibrium with B though it is separated by an adiabatic wall (*d*). It is thus equivalent to systems A and B being coupled to each other via diathermal wall (*f*), and decoupled from C by adiabatic wall (*g*) as shown in Fig. 2.1 (*b*).

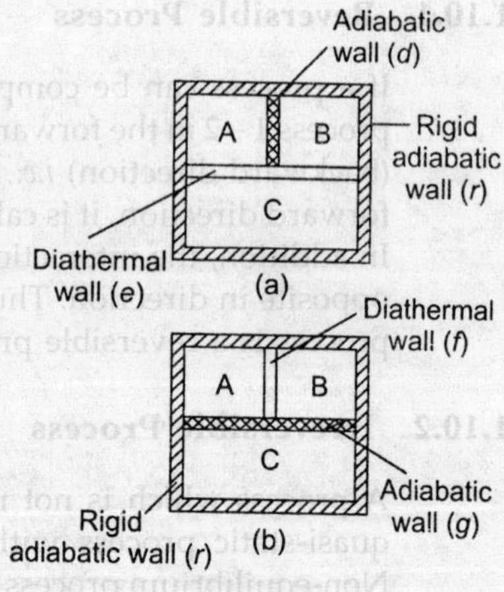

Fig. 2.1

It is to be noted that the zeroth law was introduced when first and second laws already existed at that time. So it was stated as the zeroth law so that it precedes the first and second laws to form a logical sequence.

2.2 TEMPERATURE AND ITS MEASUREMENT

The zeroth law of thermodynamics is the basis of temperature measurement. In order to obtain a quantitative measure of temperature, a reference body of known physical characteristic which changes with temperature is choosen. The choosen characteristic is called thermometric property and the reference body is called thermometer. In common thermometer, expansion of mercury is used as thermometric property. Table 2.1 shows the types of thermometer with its thermometric property.

Table 2.1

S. No	*Thermometer*	*Thermometric property*	*Symbol*
1.	Mercury-in-glass	Length	L
2.	Constant volume gas	Pressure	P
3.	Electrical resistance	Resistance	R
4.	Constant pressure gas	Volume	V
5.	Thermocouple	Thermal Electromotiveforce	E
6.	Pyrometer (Radiation)	Intensity of radiation	I or J.

2.2.1 Empirical Relations between Temperature and Thermometric Property

If x is the thermometric property and temperature is the linear function (direct proportionality) of x. So, mathematically

$$f(x) = ax$$

where a is an arbitrary constant. ...(2.1)

If x_1 corresponds to $f(x_1)$, then x_2 can be calculated as follows.

$$f(x_1) = ax_1$$

or $a = \dfrac{f(x_1)}{x_1}$ inserting this in equation (2.1) we get

$$f(x) = \frac{f(x_1)}{x_1}.x \quad ...(2.2)$$

for $x = x_2$ we get from (2.2)

$$f(x_2) = \frac{f(x_1)}{x_1}.x_2$$

or

$$\frac{f(x_2)}{f(x_1)} = \frac{x_2}{x_1} \quad ...(2.3)$$

Equation (2.3) shows that two temperatures on the linear x scale are to each other as the ratio of the corresponding readings *i.e.*, x_2 and x_1.

2.2.2 Method Used Before 1954

Let $f(x) = ax + b$ ...(2.4)

The thermometer is first placed in contact with a standard system and measures the temperature of it. Let x_1 is the reading, so,

$$f(x_1) = ax_1 + b \quad \text{...(2.5)}$$

Then the thermometer is placed with an another standard system and measures the temperature of it and let x_2 is the reading , so,

$$f(x_2) = ax_2 + b \quad \text{...(2.6)}$$

Solving (2.5) and (2.6) for a and b,

$$f(x_1) - f(x_2) = a\,(x_1 - x_2)$$

or

$$a = \frac{f(x_1) - f(x_2)}{x_1 - x_2}$$

and (2.5) $\Rightarrow$ $b = f(x_1) - ax_1$

$$= f(x_1) - \frac{f(x_1) - f(x_2)}{x_1 - x_2}\, x_1$$

$$= \frac{x_1 f(x_1) - x_2 f(x_1) - x_1 f(x_1) + x_1 f(x_2)}{x_1 - x_2}$$

or

$$b = \frac{x_1 f(x_2) - x_2 f(x_1)}{x_1 - x_2}.$$

Inserting the value of a and b in (2.4), we get

$$f(x) = \frac{f(x_1) - f(x_2)}{x_1 - x_2}\, x + \frac{x_1 f(x_2) - x_2 f(x_1)}{x_1 - x_2} \quad \text{...(2.7)}$$

Note : If $b = 0$ then equation (2.7) reduces to

$$f(x) = \frac{f(x_1) - f(x_2)}{x_1 - x_2}\, x$$

and we have

$$b = 0 = \frac{x_1 f(x_2) - x_2 f(x_1)}{x_1 - x_2}$$

or

$$x_1 f(x_2) = x_2 f(x_1)$$

or

$$\frac{f(x_2)}{f(x_1)} = \frac{x_2}{x_1} \text{ is same as equation (2.3)}$$

Two commonly used scales are Celsius scale and Fahrenheit scale. Symbols C and F are used to denote the readings on the two scales. Before 1954, the two fixed points used are steam point (boiling point of water at standard atmospheric pressure and the ice point (freezing point of water)

Temperature	*Celsius scale*	*Fahrenheit scale*
Steam point	100	212
Ice point	0	32
Interval	100 degrees	180 degrees

$$\frac{100-0}{C-0} = \frac{212-32}{F-32}$$

or $$\frac{100}{C} = \frac{180}{F-32}$$

or $$\frac{5}{C} = \frac{9}{F-32}$$

or $$\frac{C}{5} = \frac{F-32}{9} \quad ...(2.8)$$

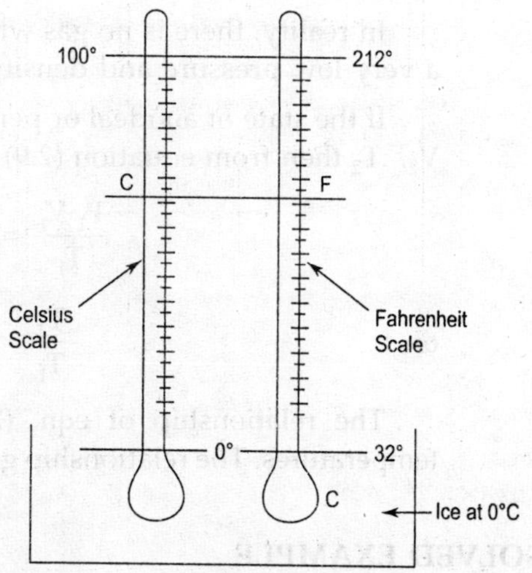

Fig. 2.2

2.2.3 Method in Use On and After 1954

On and after 1954, Kelvin suggested that only one fixed point is necessary to establish a temperature scale. This fixed point is the triple point of water. The value of the triple point of water is 0.01ºC or 273.16 K. Correspondingly, the ice point of 0ºC on the Celsius scale becomes 273.15 K on Kelvin scale. Let us use tºC for Celsius scale and T for Kelvin scale.

Then

$$\text{T in Kelvin} = 273.15 + t\text{ºC}.$$

The triple point of water in the Fahrenheit scale is 32.02ºF and for Rankine scale is 491.69R

$$\text{T in Rankine scale} = 459.67 + t\text{ ºF}$$

Table 2.2

Fixed point	*Celsius*	*Kelvin*	*Fahrenheit*	*Rankine*
Steam point	100	373.15	212	671.67
Triple point of water	0.01	273.16	32.02	491.69
Ice point	0	273.15	32	491.67
Absolute zero	– 273.15	0	– 459.67	0

2.3 IDEAL GAS

An ideal gas is one which follows the following simple equation.

$$PV = mRT \quad ...(2.9)$$

where P, V and T are the pressure, volume and temperature of the gas having mass m, and R is the gas constant. Dividing both sides of equation (2.9) by m we get,

$$P\left(\frac{V}{m}\right) = RT$$

or $$Pv = RT \text{ where } v = \frac{V}{m} = \text{specific volume of the gas.}$$

In reality, there is no gas which is an ideal gas or perfect gas. However all gases, at a very low pressure and density approach to ideal gas.

If the state of an ideal or perfect gas having mass m is changed from P_1, V_1, T_1 to P_2, V_2, T_2 then from equation (2.9) we get

$$\frac{P_1 V_1}{T_1} = \frac{P_2 V_2}{T_2} = mR = \text{constant}$$

or

$$\frac{T_2}{T_1} = \frac{P_2 V_2}{P_1 V_1} \quad \text{...(2.10)}$$

The relationship of eqn. (2.10) may be used for calculating or comparing the temperatures. The relationship given in equation (2.9) is called ideal gas equation of state.

SOLVED EXAMPLE

Example 2.1. *The temperature t (°c) on a thermometric scale is defined in terms of a property K by the relation*

$$t = a \ln K + b.$$

where a and b are constants. The values of K are found to be 1.83 and 6.78 at the ice point and the steam point, the temperature of which are assigned 0 and 100 respectively. Calculate the temperature corresponding to a reading of K equal to 2.42 on the termometer.

Solution.

$$t = a \ln K + b \quad \text{... (1)}$$

$$\left.\begin{matrix} k = 1.83 \\ t = 0 \end{matrix}\right\} \left.\begin{matrix} k = 6.78 \\ t = 100 \end{matrix}\right\}$$

$$0 = a \ln 1.83 + b \quad \text{... (2)}$$

$$100 = a \ln 6.78 + b \quad \text{... (3)}$$

Solving (2) and (3) for a and b we get

$$a = 76.355, \quad b = -46.14$$

(1) $\Rightarrow$ $\therefore$ $t = 76.355 \ln K - 46.14$

for $K = 2.42 \Rightarrow t = 21.34°C$. **Ans.**

Chapter 3

Heat and Work

3.1 INTRODUCTION

Thermodynamics deals with energy transfer in the form of heat and mechanical work. Energy can neither be created nor destroyed. It is conserved. Heat and work are energies in transition. Internal energy of a system is the storage energy. In thermodynamics we are concerned about change in energy rather than absolute energy. Hence any arbitrary state can be assigned as datum of the energy or zero level of energy.

3.2 WORK

3.2.1 Concept of Work in Mechanics

Work is done by a force when the application of force on a body moves the body in the direction of force. Work in measured by the product of the force and the distance moved in the direction of force.

The work, $W = F.d$

where F = force, d = displacement

The S.I unit of work = N.m. or Joule

1Nm = 1 J.

The action of a force through a distance (or of a torque through an angle) is called *mechanical work.*

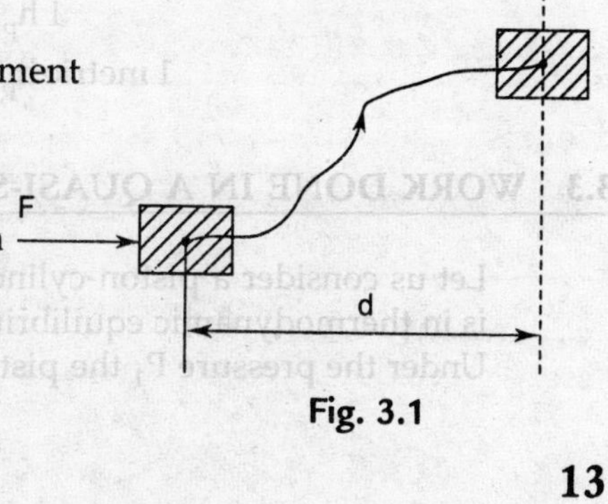

Fig. 3.1

3.2.2 Thermodynamics Concept of Work

Work is performed when the system changes its state *i.e.,* a process is done. In many thermodynamic process, the force and/or the displacement may not be visible. The definition of work needs re-defining for thermal system. W work is said to be done by a system if the sole effect on things external to the system can be reduced to the raising of a mass against gravity. The weight (*mg*) may not actually be raised, but the net effect external to the system would be the raising of a weight.

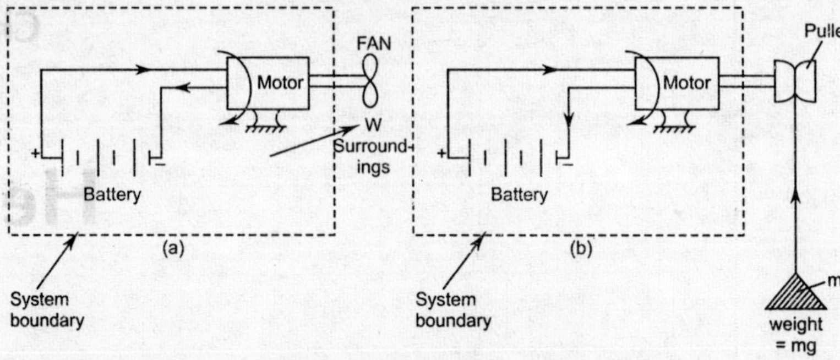

Fig. 3.2

Fig. 3.2 (*a*) shows a motor driving a fan. The system is doing work W on the surroundings (*fan*). In Fig. 3.2 (*b*) the fan is replaced by a pulley and a weight. The weight may be raised by the pulley driven by the motor.

The sole effect on things external to the system is nothing but raising the weight.

3.2.3 Convention of Sign of Work

The work is positive when it is done by the system. The work is negative when it is done on a system.

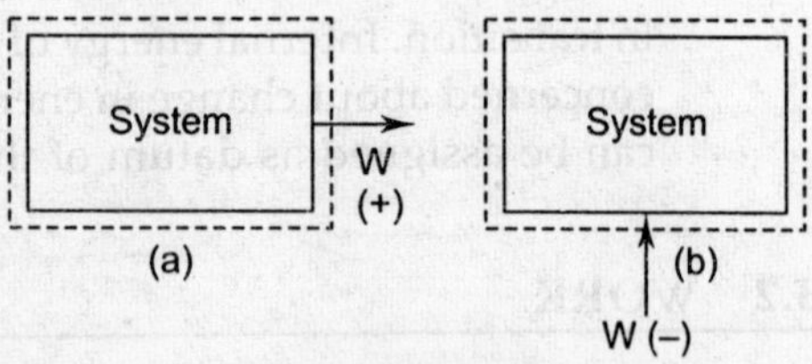

Fig. 3.3

Fig. 3.3 (*a*) shows positive work whereas Fig. 3.3 (*b*) shows negative work. Power is defined as the rate at which work is done by, or upon, the system. Thus the unit of power is 1 $^J/_S$ or watt. The system and surroundings can interact with each other through this form of energy *i.e.,* work. The unit of power is also expressed as h_p.

$$1\ h_p = 746 \text{ watt (British)}$$

$$1 \text{ metric } h_p = 745.5 \text{ watt.}$$

3.3 WORK DONE IN A QUASI-STATIC PATH

Let us consider a piston-cylinder with gas mechanism as shown in Fig. 3.4 (*a*). The gas is in thermodynamic equilibrium at position (1) which is described by the state (P_1, V_1). Under the pressure P_1 the piston moves slowly to position (2) which is also equilibrium

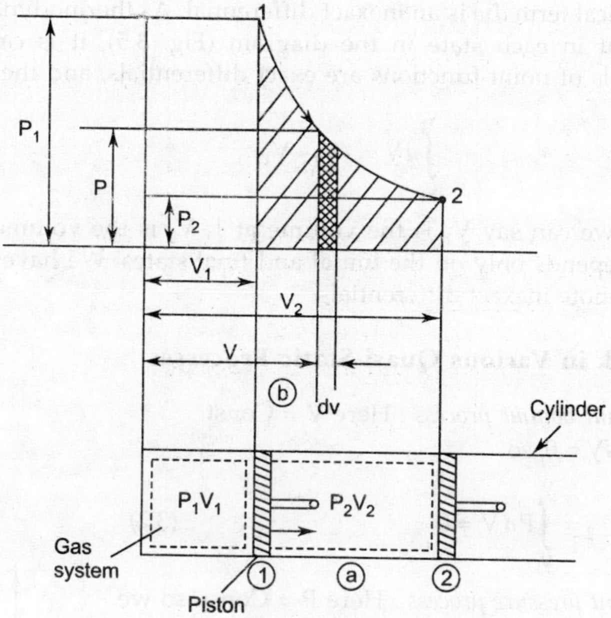

Fig. 3.4

state described by (P_2, V_2). Any intermediate stage is also in equilibrium. And at any intermediate point (P, V), the piston moves further by infinitesimal distance *dl*. If A is the area of cross section of the cylinder, the volume displaced is A*dl* = *d*V. The small work done is F.*dl*.

i.e., $$dW = F.dl = PAdl = PdV.$$

The volume changes from V_1 to V_2, therefore the total workdone is given by

$$\int_{V_1}^{V_2} đw = \int_{V_1}^{V_2} P\,dV$$

or $$W_{1-2} = \int_{V_1}^{V_2} P\,dV \qquad ...(3.1)$$

The magnitude of work can be calculated also from the area under the process (quasi-static) 1–2 as shown in Fig. 3.4 (*b*) by shaded area.

3.3.1 Path Function and Point Function

Let A, B, and C all are quasi-static paths to reach to state 2 from state 1. As per equation (3.1), the work is the area under each quasi-static process. Since the paths are different the area under each path will be different. So it is evident that the amount of workdone during each process not only is a function of the end states of the process, but also depends on the path that is followed in going from one state to another. For this reason work is called a path function or in

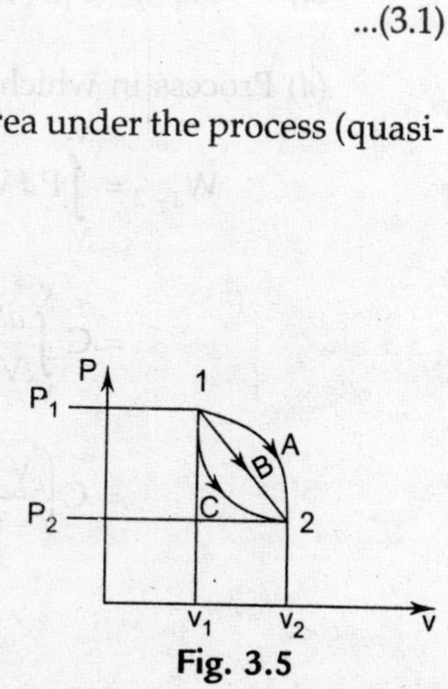

Fig. 3.5

methematical term $đw$ is an inexact differential. As thermodynamic property is having a fixed point in each state in the diagram (Fig. 3.5), it is called point function. The differentials of point functions are exact differentials, and the integration is simply

$$\int_{V_1}^{V_2} dV = (V_2 - V_1)$$

Thus we can say V_1 is the volume at 1, V_2 is the volume at 2, and the change in volume depends only on the initial and final states. We have used $đ$, (*a* cut on the tip of *d*) to denote inexact differentials.

3.3.2 PdV Work in Various Quasi-Static Processes

(*a*) *Constant volume process* : Here V = Const.
thus, $dV = 0$, so

$$W_{1-2} = \int_{V_1}^{V_2} P\,dV = 0 \qquad ...(3.2)$$

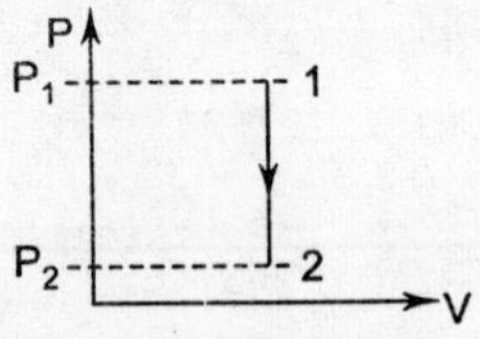

Fig. 3.6

(*b*) *Constant pressure process* : Here P = Const. so we can take out P from integration.

$$W_{1-2} = \int_{V_1}^{V_2} P\,dV$$

$$= P\int_{V_1}^{V_2} dV = P(V_2 - V_1) \qquad ...(3.3)$$

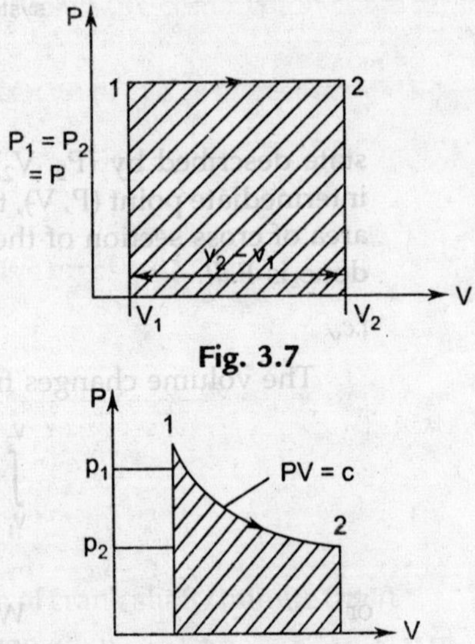

Fig. 3.7

(*c*) Process in which PV = const. = *c*.

$$W_{1-2} = \int_{V_1}^{V_2} P\,dV = \int_{V_1}^{V_2} \frac{c}{V}\,dV = c\,ln\frac{V_2}{V_1}$$

$$\text{or,} \quad W_{1-2} = P_1V_1 \ln\frac{V_2}{V_1} = P_1V_1 \ln\frac{P_1}{P_2} \qquad ...(3.4)$$

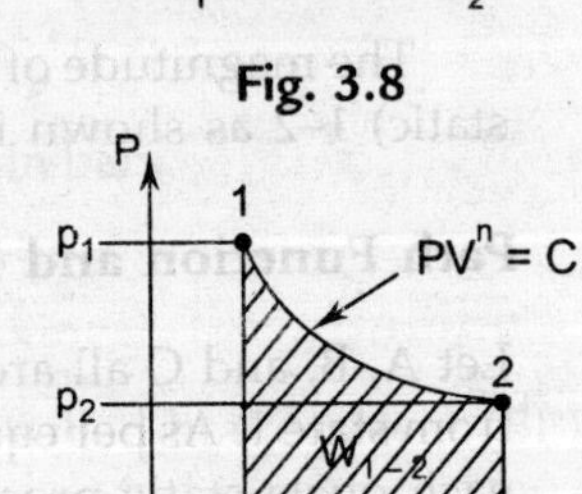

Fig. 3.8

(*d*) Process in which PV^n = Const. = C

$$W_{1-2} = \int_{V_1}^{V_2} P\,dV = \int_{V_1}^{V_2} \frac{C}{V^n}\,dV$$

$$= C\int_{V_1}^{V_2} \frac{dV}{V^n} = C\int_{V_1}^{V_2} V^{-n}\,dV$$

$$= C\left(\frac{V^{-n+1}}{-n+1}\right)_{V_1}^{V_2} = \frac{C}{1-n}\left[V_2^{1-n} - V_1^{1-n}\right] \text{ [for } n \neq 1]$$

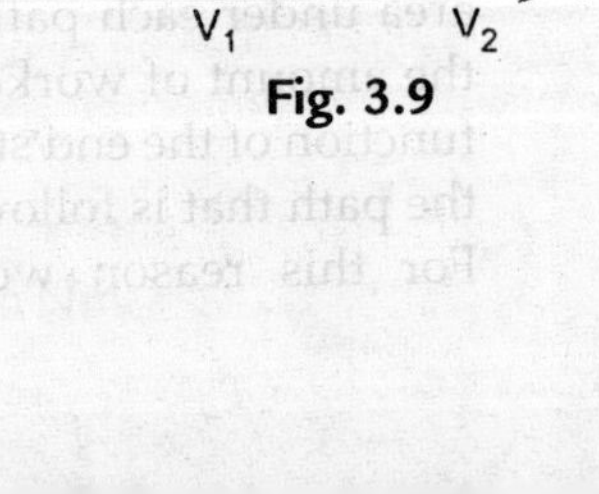

Fig. 3.9

$$= \frac{1}{1-n}\left[C V_2^{1-n} - C V_1^{1-n}\right]$$

$$= \frac{1}{1-n}\left[P_2 V_2^n V_2^{1-n} - P_1 V_1^n V_1^{1-n}\right], \qquad [C = P_1 V_1^n = P_2 V_2^n]$$

or

$$W_{1-2} = \frac{1}{1-n}\left[P_2 V_2 - P_1 V_1\right] \qquad [\text{for } n \neq 1]$$

$$= \frac{P_1 V_1 - P_2 V_2}{n-1} = \frac{P_1 V_1}{n-1}\left[1 - \left(\frac{P_2}{P_1}\right)^{\frac{n-1}{n}}\right] \qquad \text{...(3.5)}$$

If the system is having ideal gas, then we have

$$\frac{P_1 V_1}{T_1} = \frac{P_2 V_2}{T_2}$$

we get

$$\frac{T_2}{T_1} = \frac{P_2 V_2}{P_1 V_1} = \left(\frac{P_2}{P_1}\right)^{\frac{n-1}{n}} = \left(\frac{V_1}{V_2}\right)^{n-1} \qquad \text{...(3.6)}$$

$$W_{1-2} = \frac{mR\,(T_2 - T_1)}{1-n}, \; [P_2V_2 = mRT_2, P_1V_1 = mRT_1] \qquad \text{...(3.7)}$$

3.4 HEAT

3.4.1 Definition of Heat

Heat is defined as the form of energy that is transferred across the boundary of a system at a given temperature to another system (or the surroundings) at a lower temperature by virtue of the difference in temperature between the two systems. Thus heat transfer occurs solely because of the temperature difference between the two systems. This definition of heat tells that a body never contains heat. Rather, heat can be felt only when it crosses the boundary. Thus heat is energy in transition. If we consider a hot block of copper as one system and the cold water in the bucket as another system; none of them contains heat (of course, they do contain energy). When the copper block is placed in the water, heat flows from copper body to the water and after sometime a thermal equilibrium is reached. At this point we no longer have heat transfer, because there is no temperature difference between the two systems.

3.4.2 Convention of Sign of Heat

The heat is positive when it flows into the system and heat flows out of a system is taken as negative.

Here the symbol Q is used to denote heat transfer during certain time interval.

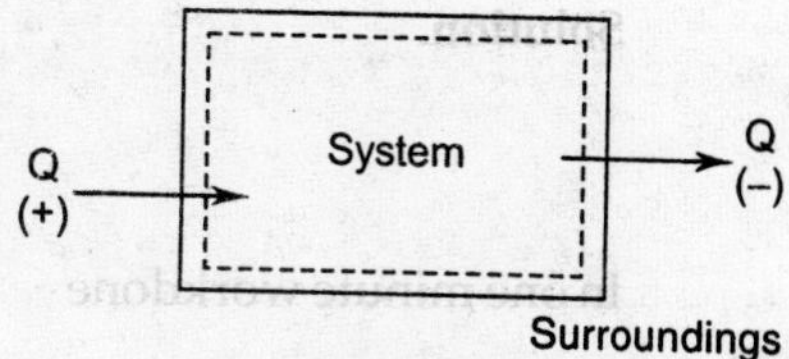

Fig. 3.10

3.4.3 Adiabatic Process

A process in which no heat crosses the boundary of the system is called *adiabatic process*. A wall which does not allow heat to cross it is called *adiabatic wall* and a wall which permits heat to cross it is called *diathermic wall.*

3.4.4 Unit of Heat

The S. I. unit of heat is joule. The rate of heat transfer or work transfer is given in watt.

3.5 HEAT TRANSFER–A PATH FUNCTION

Heat transfer is a path function. The amount of heat transfer not only depends on end states but also depends on the path followed by it. Thus we can write $đQ$ for heat transfer. It is an inexact differential. So

$$\int_1^2 đQ = Q_{1-2} \text{ or } {}_1Q_2.$$

Whenever there is temperature difference, there will be heat flow. The temperature difference is the cause and transfer of heat is the effect.

The heat transfer is written mathematically by $Q_{1-2} = \int_1^2 đQ = \int_1^2 T\,dX$

where T is the intensive property and X is the extensive property of the substance. It is valid for a quasi-static process only. The amount of heat transfer can be calculated from the area under (T, X) coordinates diagram

$$đQ = T\,dX$$

or $$dX = \frac{đQ}{T}$$

here $\frac{1}{T}$ is called integrating factor used with inexact quantity to make it exact differentials.

SOLVED EXAMPLES

Example 3.1. *A centrifugal pump forces 2 m³/min of water horizontally from an open well to a closed tank, where the pressure is 0.45 MPa. Compute the work the pump must do upon the water in an hour just to force the water into the tank against the pressure.*

Solution. $P = 0.45 \times 10^6\ Pa = 450\ KPa$

$dV = 2m^3$, time, $\tau = 1$ min

$= V_2 - V_1$

In one minute workdone $= \int P\,dV = P(V_2 - V_1)$

$= 450\,(2 - 0) = 900$ KNm.

Thus in one hour the workdone $= 900 \times 60$ KJ/hr.

$= 54000$ KJ/hr.

Example 3.2. *A petrol engine cylinder has a piston of cross section 0.15 m^2. It contains gas at a pressure of 2MPa. The gas expands according to a straight line process on P – V diagram. The final pressure is 0.2 Mpa. Calculate the workdone by the gas on the piston if the stroke is 0.15 m.*

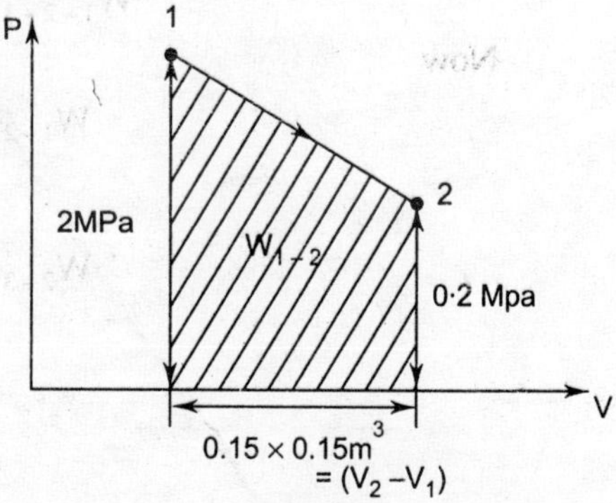

Fig. 3.11

Solution.

W_{1-2} = area under the curve (Straight line) 1 – 2

$$= \frac{1}{2}[2 + 0.2] \times 10^6 \times 0.15 \times 0.15 \text{ Nm}$$

$= 24.75$ KJ. **Ans.**

Example 3.3. *A mass of 2 kg of air is compressed in a quasi-static process from 0.15 MPa to 0.75 MPa for which PV = Const. The initial density of air is 1.5 kg/m^3. Calculate the workdone by the piston to compress the air.*

Solution. We have

$$W_{1-2} = P_1 V_1 \ln \frac{P_1}{P_2}$$

$$= 0.15 \times 10^6 \times \frac{2}{1.5} \ln \frac{0.15}{0.75} \text{ J}$$

$$= -0.321888 \times 10^6 \text{ J}$$

$$= -321.8 \text{ KJ}.$$

The negative sign of work shows that the work is done on the system.

Example 3.4. *A quantity of gas is compressed in a quasi-equilibrium process from 90 KPa, 0.15 m^3 to 0.5 MPa, 0.04 m^3. Assuming that the pressure and volume are related by PV^n = Const., calculate the workdone on the gas system. (Take n = 1.3)*

Solution.

$$W_{1-2} = \frac{P_1 V_1 - P_2 V_2}{n-1} = \frac{90 \times 10^3 \times 0.15 - 0.5 \times 10^6 \times 0.04}{1.3 - 1} \text{ J}$$

$= -21.7$ KJ.

Negative sign shows that the work is done on the system.

Example 3.5. *Calculate the total workdone by a gas system in a expansion process shown in Fig. 3.12 (1 bar = 10^5 N/m^2)*

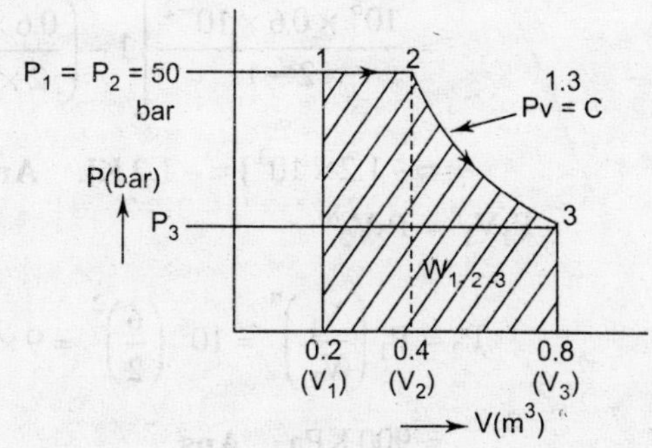

Fig. 3.12

Solution.

$$W_{1-2-3} = W_{1-2} + W_{2-3}$$

Now

$$W_{1-2} = 50 \times 10^5 \times (0.4 - 0.2)\ \text{J} = 10 \times 10^5\ \text{J} \quad \textbf{[Note. } P_2 V_2^n = P_3 V_3^n\textbf{]}$$

$$W_{2-3} = \frac{P_2 V_2 - P_3 V_3}{n-1} = \frac{P_2 V_2}{n-1}\left[1 - \left(\frac{P_3}{P_2}\right)^{\frac{n-1}{n}}\right]$$

$$= \frac{P_2 V_2}{n-1}\left[1 - \left(\frac{V_2}{V_3}\right)^{n-1}\right]$$

$$= \frac{50 \times 10^5 \times 0.4}{1.3 - 1}\left[1 - \left(\frac{0.4}{0.8}\right)^{0.3}\right]$$

$$= 12.516 \times 10^5\ \text{J}.$$

$$W_{1-2-3} = (10 \times 10^5 + 12.516 \times 10^5)\ \text{J}$$

$$= 22.516 \times 10^5\ \text{J}$$

$$= 2.2516 \times 10^6\ \text{J} = 2.2516\ \text{MJ}. \quad \textbf{Ans.}$$

Example 3.6. *If a gas of volume $0.6 \times 10^{-2}\ m^3$ and at pressure $10^5\ N/m^2$ is compressed quasi-statically according to PV^2 = const. until the volume becomes $2 \times 10^{-3}\ m^3$, calculate the final pressure and work transfer.*

Solution.

$$W_{1-2} = \frac{P_1 V_1 - P_2 V_2}{n-1}$$

$$= \frac{P_1 V_1}{n-1}\left[1 - \left(\frac{V_1}{V_2}\right)^{n-1}\right]$$

$$= \frac{10^5 \times 0.6 \times 10^{-2}}{2-1}\left[1 - \left(\frac{0.6 \times 10^{-2}}{2 \times 10^{-3}}\right)^{2-1}\right]$$

$$= -1.2 \times 10^3\ \text{J} = -1.2\ \text{KJ}. \quad \textbf{Ans.}$$

$$P_1 V_1^n = P_2 V_2^n$$

or

$$P_2 = P_1\left(\frac{V_1}{V_2}\right)^n = 10^5\left(\frac{6}{2}\right)^2 = 9 \times 10^5\ N/m^2$$

$$= 900\ \text{KPa}. \quad \textbf{Ans.}$$

Chapter 4

First Law of Thermodynamics (Control Mass/Closed System)

4.1 FIRST LAW OF THERMODYNAMICS FOR A CONTROL MASS (CLOSED SYSTEM) UNDERGOING A CYCLE

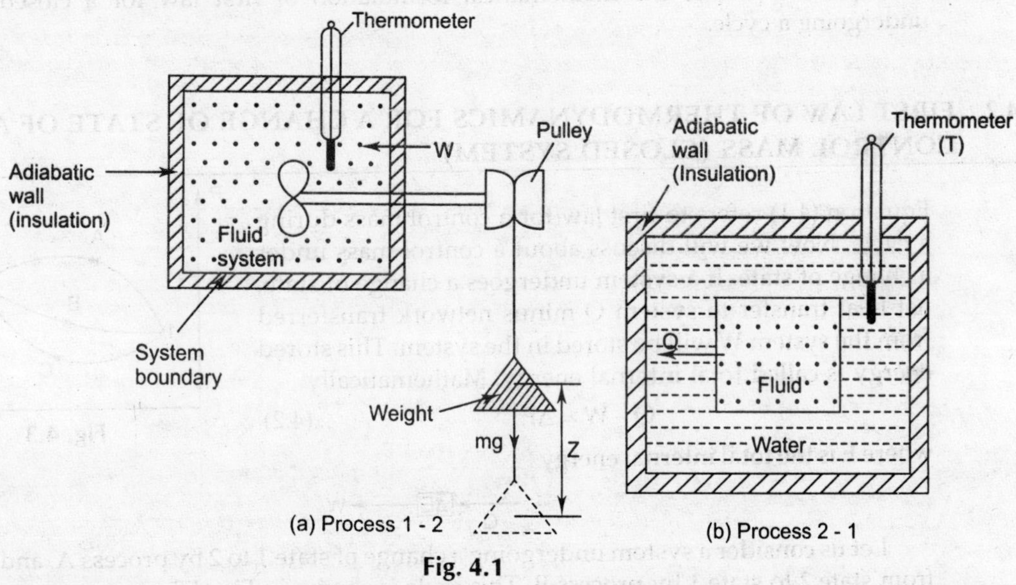

Fig. 4.1

J.P. Joule conducted his experiments during 1840–1849. His apparatus and experiments are shown in Fig. 4.1.

He took a fluid system (Fig. 4.1 (*a*)) and work was done on the system by means of paddle wheel. The amount of work W = mgz. The temperature of the fluid system was increased. Now the system was placed in a water bath as shown in Fig. 4.1 (*b*). Heat Q was transferred to the water bath as seen through thermometer until the original state of the fluid was reached. The amount of heat transferred was measured from the increase in energy of water.

He then repeated the experiments by taking different systems and surroundings. Joule found that ratio of workdone on the system to heat generated was a constant. Mathematically.

$$\frac{\oint đW}{\oint đQ} = J = a \text{ constant called Joule's constant}$$

or $$\oint đW = J \oint đQ$$

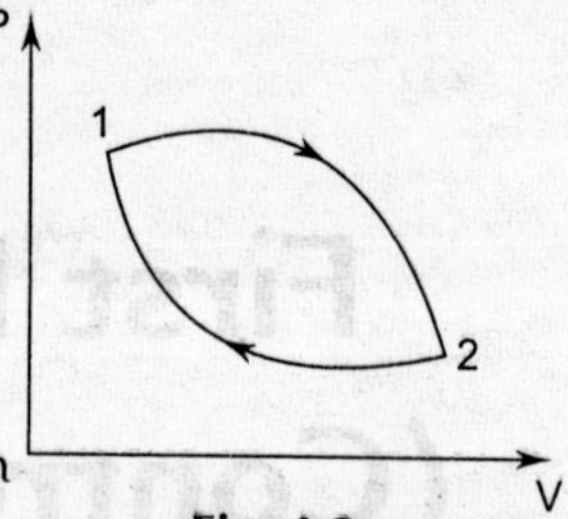

Fig. 4.2

If work and heat is expressed in the same unit (Joule) then 1J = 1Nm.

So $$\oint đW = \oint đQ \qquad ...(4.1)$$

[symbol $\oint$ denotes the cyclic integral for closed path]

Now we can define first law of thermodynamics as "In a closed system undergoing cyclic operations, the net work transfer is equal to net heat transfer".

Equation (4.1) is the mathematical formulation of first law for a closed system undergoing a cycle.

4.2 FIRST LAW OF THERMODYNAMICS FOR A CHANGE OF STATE OF A CONTROL MASS (CLOSED SYSTEM)

Equation (4.1) refers to first law for a control mass during a cycle. Now we will discuss about a control mass under a change of state. If a system undergoes a change of state, net heat transfer to system Q minus network transferred from the system W will be stored in the system. This stored energy is called total internal energy. Mathematically

$$Q - W = \Delta E \qquad ...(4.2)$$

where E is the total internal energy

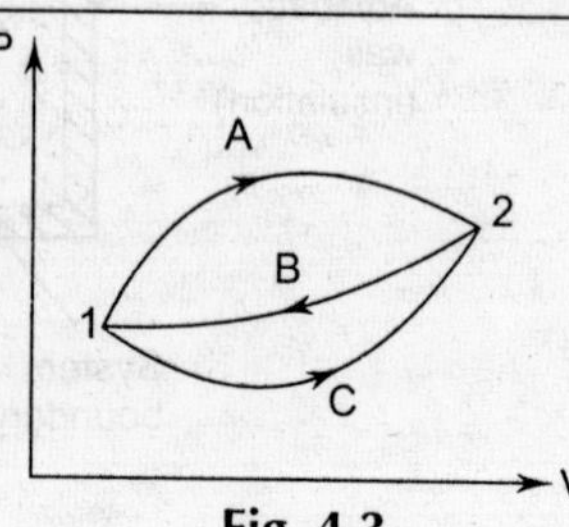

Fig. 4.3

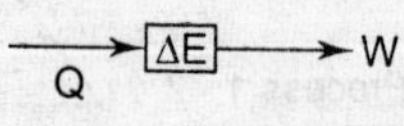

Let us consider a system undergoing a change of state 1 to 2 by process A, and returns from state 2 to state 1 by process B. This cycle is shown in Fig. 4.3 on a pressure-volume coordinate system (any other intensive and extensive property can be used as a coordinate system).

From equation (4.1) we can write

$$\oint đQ = \oint đW$$

Considering processes A and B together we get

$$\int_1^2 đQ_A + \int_2^1 đQ_B = \int_1^2 đW_A + \int_2^1 đW_B \qquad ...(4.3)$$

Now we consider another cycle in which the control mass changes from state 1 to state 2 by process C and returns back to state 1 by process B. For this cycle we can write,

$$\int_1^2 đQ_C + \int_2^1 đQ_B = \int_1^2 đW_C + \int_2^1 đW_B \qquad ...(4.4)$$

Subtracting eqn. (4.4) from (4.3) we get

$$\int_1^2 đQ_A - \int_1^2 đQ_C = \int_1^2 đW_A - \int_1^2 đW_C$$

or

$$\int_1^2 đQ_A - \int_1^2 đW_A = \int_1^2 đQ_C - \int_1^2 đW_C$$

$$\int_1^2 \left(đQ - đW\right)_A = \int_1^2 \left(đQ - đW\right)_C$$

Since A and B are two arbitrary processes between states 1 and 2, the quantity ($đQ - đW$) is the same for all processes between states 1 and 2. Therefore ($đQ - đW$) depends on end states (here 1 and 2) and not by the path it traces. So this is a point function. Therefore it is an exact differentials. This property is the energy of the mass and call total internal energy. Thus mathematically

$$đQ - đW = dE$$

or

$$đQ = dE + đW \qquad ...(4.4(a))$$

Integrating between states 1 and 2, we get

$$Q_{1-2} = (E_2 - E_1) + W_{1-2}$$

where E_1 and E_2 are the value of the total internal energy of the control mass in state 1 and 2 respectively. Q_{1-2} is the heat supplied during 1 to 2 and W_{1-2} is the work output from the system between states 1 – 2.

4.3 DIFFERENT TYPES OF STORED ENERGY

E represents the total energy stored in a system. There are two modes in which the energy may be stored in a system.

- Macroscopic energy mode,
- Microscopic energy mode.

4.3.1 Macroscopic Energy Mode

Kinetic energy and potential energy of a system are macroscopic energy.

Kinetic Energy = $E_k = \frac{1}{2} m\overline{V}^2$ where m is the mass of the system and $\overline{V}$ is the centre of mass velocity and,

Potential Energy = $E_p = mgz$ where z is the elevation or height from the datum. This is the energy stored by virtue of position of the system.

4.3.2 Microscopic Energy Mode

Molecular internal energy or simply internal energy (U) stored in molecules or atoms of a system is the microscopic energy. A system is composed of matter which contains molecules and atoms. Molecules of gases are in random motion with an average velocity $\vec{v}$. They collide with other molecules and walls of the container and due to collision, the molecules may gain rotational energy, and vibrational energy. They also translational, electronic, chemical, nuclear energy. If $\in$ represents the total energy of one molecule, then

$$\in = \in_{trans} + \in_{rot} + \in_{vibra} + \in_{chem} + \in_{electronic} + \in_{nuclear}$$

If N is the total number of molecules in a system then total internal energy U is given by

$$U = N\in \quad ...(4.5)$$

For ideal gas, U is a function of T only

$$U = f(T) \quad ...(4.6)$$

Magnetic energy, electrical energy and surface tension energy can also be present in the system. In absence of these energies, the total stored energy can be written as

$$E = E_K + E_P + U$$

In absence of motion and gravity

i.e.,
$$E_K = 0 \text{ and } E_P = 0,$$
$$E = U$$

So equation (4.4 (*a*)) becomes.

$$đQ = dE + đW$$

or
$$đQ = dU + đW \quad ...(4.7)$$

In absence of shaft work, electrical work etc., we can write

$$đW = \text{Displacement work only}$$
$$= PdV$$

or from (4.7) $đQ = dU + PdV$. Integrating between state 1 and 2 we get

$$\int_1^2 đQ = \int_1^2 dU + \int_1^2 PdV$$

or
$$Q_{1-2} = U_2 - U_1 + \int_1^2 PdV$$

$$Q_{1-2} = U_2 - U_1 + W_{1-2} \quad ...(4.8)$$

If we consider Kinetic Energy and potential energy equation (4.8) can be expanded to

$$Q_{1-2} = U_2 - U_1 + \underbrace{\frac{1}{2}m\left(\bar{V}_2^2 - \bar{V}_1^2\right)}_{\text{KE change}} + \underbrace{mg\,(z_2 - z_1)}_{\text{PE change}} + W_{1-2} \quad ...(4.9)$$

Note that internal energy is an extensive property because it depends on the mass of the system. Similarly Kinetic Energy and Potential Energy are extensive properties of a system.

Specific internal energy (u) is defined as the internal energy per unit mass, so.

$$u = \frac{U}{m}.$$

4.4 THE CONSTANT-VOLUME SPECIFIC HEAT

We have

$$đQ = dU + đW$$

$$= dU + PdV$$

If the volume is constant *i.e.*, V = C or $dV = 0$

$$\text{So } đW = PdV = 0$$

or
$$(đQ)_v = (dU)_v \qquad \text{...(4.10)}$$

Specific heat is defined as the amount of heat required to raise the temperature of unit mass by one degree.

So the specific heat (c_v) is

$$c_v = \frac{1}{m}\left(\frac{đQ}{dT}\right)_V = \frac{1}{m}\left(\frac{dU}{dT}\right)_V$$

$$= \left(\frac{\partial u}{\partial T}\right)_V \quad \left[\text{as } \frac{dU}{m} = \partial u\right] \qquad \text{...(4.11)}$$

So we can properly redefine the specific heat at constant volume as "It is the rate of change of specific internal energy with respect to temperature when the volume of system is kept constant".

We also have
$$Q_{1-2} = U_2 - U_1 + W_{1-2}$$

$$= m(u_2 - u_1) + W_{1-2}$$

or
$$\frac{Q_{1-2}}{m} = u_2 - u_1 + \frac{W_{1-2}}{m}$$

or
$$q_{1-2} = u_2 - u_1 + w_{1-2}$$

[Lower case is used to define specific quantity]

or
$$đq = du + đw.$$

$$(du)_V = c_v\, dT$$

or
$$u = \int_{T_1}^{T_2} c_v\, dT$$

Equation (4.10) states that heat transfer at constant volume increases the internal energy of the system.

Equation (4.11) is a proper definition of c_v rather than the previous one.

Heat capacity at constant volume is the product of mass to the specific heat at constant volume.

$\therefore$ $mc_v = C_v$. The unit of $C_v = J/°K$.

4.5 ENTHALPY

We have $$Q_{1-2} = U_2 - U_1 + W_{1-2} \quad ...(4.12)$$

$$W_{1-2} = \int_1^2 P\,dV, \text{ If P is constant}$$

$$= P\int_1^2 dV = P(V_2 - V_1)$$

or $$= P_2V_2 - P_1V_1 \quad [P_2 = P_1 = P]$$

(4.12) $\Rightarrow$ $$Q_{1-2} = U_2 - U_1 + P_2V_2 - P_1V_1$$

$$= (U_2 + P_2V_2) - (U_1 + P_1V_1) \quad ...(4.13)$$

Equation (4.13) shows a very restricted case where the heat transfer during the process is given in terms of the change in the quantity (U + PV) between the state 1 and 2. Each term of (U + PV) is the property of a substance that is a function of the state of the system, their combination must also be a property. We term this property as enthalpy. So

$$H \text{ (enthalpy)} = U + PV$$

or $$\frac{H}{m} = \frac{U}{m} + \frac{PV}{m}$$

or $$h = u + Pv \quad ...(4.14)$$

where h = specific enthalpy; the S.I. unit of h is J/kg.

We have $$đq = du + đw$$

$$= du + P dv$$

$$= du + d(Pv)$$

$$= d(u + Pv)$$

$$= dh.$$

4.6 THE CONSTANT-PRESSURE SPECIFIC HEAT

The specific heat at constant pressure is defined as the rate of change of specific enthalpy with respect to temperature.

$$c_p = \left(\frac{\partial h}{\partial T}\right)_P$$

For constant pressure process

$$đq = dh$$

or $$dh = \int_{T_1}^{T_2} c_p\, dT.$$

4.7 SPECIFIC HEATS OF SOLIDS AND LIQUIDS

c_p and c_v are used particularly in case of gases. In case of solids and liquids, the specific volume is very small and its changes can be neglected.

Thus $h = u + Pv$

or $dh = du + d\,(Pv)$

$= du + 0 \quad \therefore \quad dh \approx du.$

Hence $c_v = c_p = c$ (say).

Thus there is only one specific heat c for solids and liquids. So in any process.

$$đq = dh = du = c dT.$$

4.8 ENERGY OF ISOLATED SYSTEM

As per the definition of isolated system $đQ = 0$ and $đW = 0$.

Thus $đQ = dE + đW$

or $0 = dE + 0$

or $dE = 0$

or $E = \text{constant}.$

Thus the total internal energy or stored energy is constant.

4.9 PERPETUAL MOTION MACHINE OF THE FIRST KIND (PMM-1)

A machine which continuously produces work from nothing is called a perpetual motion machine of the first kind. As per Fig. 4.4, W (work) is produced without taking any energy from surroundings. It creats its own energy. This concept violates the first law of thermodynamic. All efforts to create such PMM-1 have failed, thus providing the experimental proof of the validity of the first law of thermodynamics.

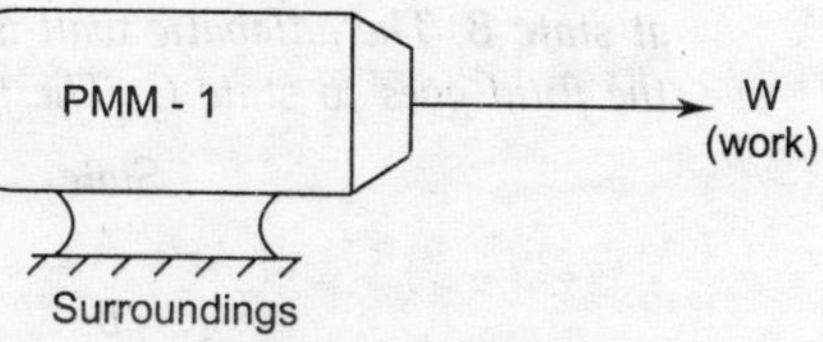

Fig. 4.4

SOLVED EXAMPLES

Example 4.1. *An engine is tested by means of a water brake at 1100 rpm. The torque of the engine is 11000 Nm and the water consumption of the brake is 1 m^3/S, its inlet tamperature being 25°C. Calculate the water temperature at exit, assuming that the whole of the engine power is ultimately transformed into heat which is absorbed by the cooling water (specific heat of water = c_p = 4.187 KJ/KgK).*

Solution. Given $T = 11000$ Nm

$N = 1100$ *rpm*

$t_1 = 25°C, t_2 = ?$

Power, $P = Tw = T \times \dfrac{2\pi N}{60}$; $\left[w = \dfrac{2\pi N}{60}\right]$

or $P = \dfrac{11000 \times 2\pi \times 1100}{60}$ J/s.

$$\text{work} = \text{heat} = P \times t$$

or $m\,c_p(T_2 - T_1) = P \times t$

or $\dfrac{m}{t}\,c_p(T_2 - T_1) = P,$ $\left[\dfrac{m}{t} = 1 \times 10^3\ \dfrac{Kg}{s}\right]$

or $1 \times 10^3 \times 4.187 \times 10^3 \times (t_2 - t_1) = \dfrac{11000 \times 2\pi \times 1100}{60}$

or $t_2 - t_1 = \dfrac{11 \times 2\pi \times 11}{60 \times 41.87} = 0.3$

or $t_2 = (t_1 + 0.3) = 25.3°C$. **Ans.**

Example 4.2. *In a cyclic process, heat transfers are + 15 KJ, – 26 KJ, – 4.56 KJ and + 32.56 KJ. Calculate the net work transfer.*

Solution.

In a cyclic operation,

$$\Sigma Q_{cycle} = \Sigma W_{cyclic}$$

or $(+15 - 26 - 4.56 + 32.56)\ KJ = \Sigma W_{cycle}$

or $17\ KJ = \Sigma W_{cycle}$ = net work transfer. **Ans.**

Example 4.3. *A slow chemical reaction takes place in a fluid at the constant pressure of 10^5 N/m². The fluid is surrounded by adiabatic wall during the reaction which begins at state A and ends at state B. The adiabatic wall is then removed and 106 KJ of heat flows to the surroundings as the fluid goes to state C. The following data are given.*

State	*V(m³)*	*t(°C)*
A	*0.004*	*20*
B	*0.4*	*370*
C	*0.08*	*20*

Calculate E_B and E_C if E_A = O for the fluid system.

Solution. For states A to B

we have $đQ = dE + đW$

or integrating from 1 – 2

$$Q_{A-B} = E_B - E_A + \int_A^B P\,dV, \quad \text{[P is constant]}$$

$$= E_2 - E_1 + P(V_2 - V_1)$$

as adiabatic wall was there, so

$$Q_{A-B} = 0 \text{ and } E_A = 0 \text{ given}$$

or $O = E_B - 0 + 10^5\,(0.4 - 0.004)$

or $E_B = -10^5\,(0.4 - 0.004)$ Joule

$$= -39.6 \times 10^3 \text{ Joule} = -39.6 \text{ KJ.} \quad \textbf{Ans.}$$

For states B to C

$$Q_{B-C} = E_C - E_B + P(V_3 - V_2) \qquad [P = \text{atmospheric pressure}]$$

$$-106 \times 10^3 = E_C + 39.6 \times 10^3 + 10^5 (0.08 - 0.4)$$

[Q_{BC} is heat out from system so it is negative]

or $$E_C = -113.6 \times 10^3 \text{ Joule}$$

or $$E_C = -113.6 \text{ KJ.} \quad \textbf{Ans.}$$

Example 4.4. *During a certain period a machine consumes 1 KWh of energy and the internal energy of the system drops by 5000 KJ. Find out the net heat transfer for the system.*

Solution. Work W_{1-2} is done on the system.

∴ $$W_{1-2} = -1 \text{ KWh} = -3600 \text{ KJ}$$

Internal energy drops *i.e.,* $E_2 - E_1 = -5000$ KJ

We know

$$Q_{1-2} = E_2 - E_1 + W_{1-2}$$

or $$= -5000 \text{ KJ} - 3600 \text{ KJ}$$

$$= -8600 \text{ KJ}$$

$$= -8.6 \text{ MJ.} \quad \textbf{Ans.}$$

Example 4.5. *The properties of a certain fluid are related as follows :*

$$u = 196 + 0.718\, t$$

and $$Pv = 0.287\,(t + 273)$$

where u is the specific internal energy (KJ/kg), t is in °C and P is pressure in KN/m², v is the specific volume (m³/kg). Calculate c_v and c_p for this fluid.

Solution. for c_v $u = 196 + 0.718\,t$ $\qquad [T = t + 273]$, $t = (T - 273)$

$$= 196 + 0.718\,(T - 273)$$

$$= 196 + 0.718\,T - 196.014$$

$$= 0.718\,T - 0.014$$

or $$\left(\frac{\partial u}{\partial T}\right)_v = 0.718 = c_v$$

∴ $$c_v = 0.718 \quad \textbf{Ans.}$$

For c_p $$h = u + Pv$$

$$= 0.718\,T - 0.014 + 0.287\,T$$

$$= 1.005\,T - 0.014$$

or $$\left(\frac{\partial h}{\partial T}\right)_p = 1.005 = c_p \quad \therefore \quad c_p = 1.005. \quad \textbf{Ans.}$$

Example 4.6. *A mass of 8 kg gas expands within a flexible container so that the P – v relationship follows $Pv^{1.2}$ = C equation. The initial pressure is 1000 KPa and the initial volume is 1 m³. The final pressure is 5 KPa. If specific internal energy of the gas decreases by 40 KJ/kg. Find the heat transfer in magnitude and direction.*

Solution.

$$W_{1-2} = \frac{P_1 v_1}{n-1}\left[1-\left(\frac{P_2}{P_1}\right)^{\frac{n-1}{n}}\right]$$

$$= \frac{1000 \times 10^3 \times 1}{1.2-1}\left[1-\left(\frac{5}{1000}\right)^{\frac{1.2-1}{1.2}}\right]$$

$$= 2932.4 \text{ KJ}$$

$$Q_{1-2} = E_2 - E_1 + W_{1-2}$$

$$q_{1-2} = e_{1-2} + w_{1-2}$$

$$= \left(-40 + \frac{2932.4}{8}\right) \text{KJ/kg}$$

$$= +326.55 \text{ KJ/kg}$$

$$Q_{1-2} = 326.55 \times 8 \text{ KJ}$$

$$= +2612.4 \text{ KJ.} \quad \textbf{Ans.}$$

Example 4.7. *A gas of mass 1.5 kg undergoes a quasi-static expansion according to P = a + b V relationship (a, b, constants). The initial pressure and volume are 1000 KPa and 0.2 m^3 respectively. The final pressure and volume are 200 KPa and 1.20 m^3. The specific internal energy of the gas is given by u = 1.5 Pv – 85 KJ/kg where pressure P in KPa and v in m^3/kg units. Calculate the net heat transfer and the maximum internal energy of the gas attained during expansion.*

Solution.

$$P = a + bV$$

at $\left.\begin{array}{l} P = 1000 \\ V = 0.2 \end{array}\right\} \Rightarrow 1000 = a + 0.2\,V$ ...(1)

at $\left.\begin{array}{l} P = 200 \\ V = 1.2 \end{array}\right\} \Rightarrow 200 = a + 1.2\,V$ ...(2)

Solving (1) and (2) we get $a = 1160$

$$b = -800$$

$\therefore$ $P = 1160 - 800\,V$ ...(3)

$$W_{1-2} = \int_{V_1}^{V_2} P\,dV = \int_{V_1}^{V_2} (1160 - 800\,V)\,dV$$

$$= \int_{0.2 = V_1}^{1.2 = V_2} (1160 - 800\,V)\,dV = 1160\,(V_2 - V_1) - 800 \times \frac{1}{2}\,(V_2^2 - V_1^2)$$

$$= 1160\,(1.2 - 0.2) - 800 \times \frac{1}{2}\,(1.2^2 - 0.2^2)$$

$$= 600 \text{ KJ.}$$

We also have $u = 1.5\,Pv - 85$

or $U = mu = 1.5\,PV - 85 \times m$

or $U = 1.5\,PV - 85 \times 1.5$ ...(4)

or $$\int_1^2 dU = \int_1^2 1.5P\,dV + \int_1^2 1.5V\,dP$$

or $$U_2 - U_1 = 1.5 \times 600 + 1.5\,(-800)\int_{V_1}^{V_2} V\,dV \qquad [dP = -800\,dV]$$

$$= 900 - \frac{1200}{2}[V_2^2 - V_1^2]$$
$$= 900 - 600\,[1.2^2 - 0.2^2]$$
$$= 900 - 840 = 60 \text{ KJ.}$$

We also have $Q_{1-2} = U_2 - U_1 + W_{1-2}$

$= (60 + 600)$ KJ $= 660$ KJ. **Ans.**

Now from eqn (4) we have

$$U = 1.5\,PV - 85 \times 1.5 \qquad \left[v = \frac{V}{m}\right]$$
$$= 1.5\,(1160 - 800V)V - 85 \times 1.5$$
$$= 1740\,V - 1200\,V^2 - 85 \times 1.5 \qquad ...(5)$$
$$\frac{dU}{dV} = 1740 - 2400\,V$$

for $U = U_{max}$, $\dfrac{dU}{dV} = 0$

or $0 = 1740 - 2400\,V \Rightarrow V = 0.725\ m^3$

Putting this value in (5) we get,

$$U_{max} = 1740 \times 0.725 - 1200 \times 0.725^2 - 85 \times 1.5$$
$$= 503.25 \text{ KJ.}$$ **Ans.**

Chapter **5**

First Law of Thermodynamics (Control Volume/Open System)

5.1 OPEN SYSTEM AND CONTROL VOLUME

In the previous chapter we have derived energy balance equation considering a control mass system. The analysis was made in a closed system. However most of the practical devices deal with open system. The mass, work and heat can enter and exit the system. We introduce concept of control volume which is a fixed region in space. The size and shape of the control volume are completely arbitrary and are so defined to best suit our analysis. The surface of this control volume is called *control surface*. Control surface is a closed surface. The surface may be fixed, or it may move to expand or contract. However, the surface must be defined relative to some coordinate system. The coordinate system may be fixed, rotating or moving-depending upon the situation under analysis.

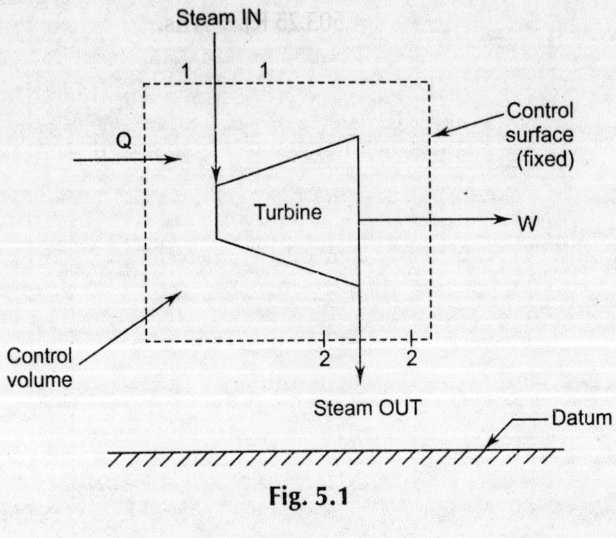

Fig. 5.1

Fig. 5.1 shows a control volume and control surface. Mass, heat and work transfers are also shown in Fig. 5.1. We have shown here fixed control surface and fixed datum.

Fig. 5.1 also shows that it is an open system.

5.2 EQUATION FOR CONSERVATION OF MASS

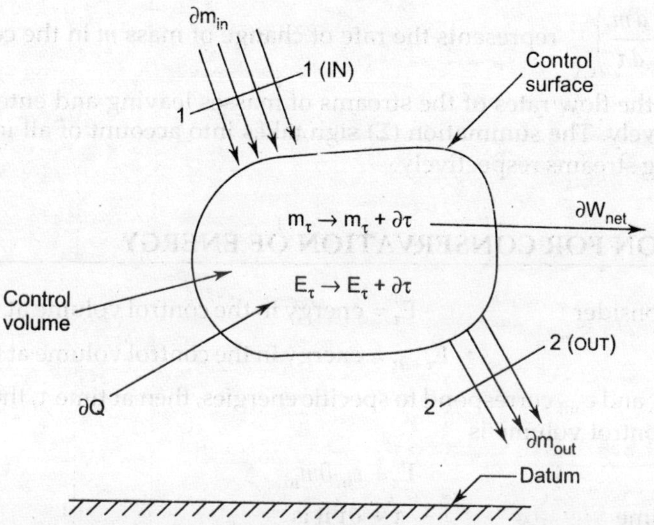

Fig. 5.2

To apply the first law of thermodynamics, we reduce the situation to a fixed or control mass problem. A general representation of an open system is shown in Fig. 5.2 where the working substance in the control volume at time τ has mass m_τ and energy E_τ. And at time interval of $\partial\tau$, mass ∂m_{in} at state 1 enters the control volume, and mass, ∂m_{out} at state 2 leaves the control volume. In addition heat supplied to the control volume is ∂Q and amount of work transfer is ∂W_{net}. As a result the mass at $\tau + \partial\tau$ is $m_{\tau+\partial\tau}$ and energy is $E_{\tau+\partial\tau}$.

We can write down $\begin{bmatrix} 1-1 \Rightarrow \text{IN section} \\ 2-2 \Rightarrow \text{OUT section} \end{bmatrix}$

$$\text{Initial mass} = m_\tau + \delta m_{in}$$

$$\text{final mass} = m_{\tau+\partial\tau} + \partial m_{out}$$

So, mass balance is given by

$$m_\tau + \partial m_{in} = m_{\tau+\partial\tau} + \partial m_{out}$$

or
$$(m_{\tau+\delta\tau} - m_\tau) + (\delta m_{out} - \delta m_{in}) = 0$$

Dividing by δ_τ and taking $\partial_\tau \to 0$, we obtain

$$\underset{\delta\tau \to 0}{\text{Lt}} \left(\frac{m_{\tau+\partial\tau} - m_\tau}{\partial\tau} \right) + \frac{\partial m_{out}}{\partial\tau} - \frac{\partial m_{in}}{\partial\tau} = 0$$

$$\left(\frac{dm}{\partial \tau}\right)_{CV} + \frac{dm_{out}}{d\tau} - \frac{dm_{in}}{d\tau} = 0. \text{ Thus ingeneral} \qquad ...(5.1)$$

or $$\left(\frac{dm}{d\tau}\right)_{CV} + \Sigma \dot{m}_{out} - \Sigma \dot{m}_{in} = 0 \text{ [CV = control volume]} \qquad ...(5.2)$$

where $\left(\frac{dm}{d\tau}\right)_{CV}$ represents the rate of change of mass m in the control volume, $\dot{m}_{out}$ and $\dot{m}_{in}$ are the flow rates of the streams of masses leaving and entering the control volume respectively. The summation (Σ) sign takes into account of all incoming streams and all outgoing streams respectively.

5.3 EQUATION FOR CONSERVATION OF ENERGY

Let us consider E_τ = energy in the control volume at time τ, and

$E_{\tau+\partial\tau}$ = energy in the control volume at time $\tau + \partial\tau$.

If e_{in} and e_{out} correspond to specific energies, then at time τ, the total energy of the mass of the control volume is

$$E_\tau + e_{in}\,\partial m_{in}$$

and at time $\tau + \partial\tau$ it is

$$E_{\tau+\partial\tau} + e_{out}\,\partial m_{out}.$$

So net energy change
$$= dE = (E_{\tau+\partial\tau} + e_{out}\,\partial m_{out}) - (E_\tau + e_{in}\,\partial m_{in})$$
$$= (E_{\tau+\partial\tau} - E_\tau) + (e_{out}\,\partial m_{out} - e_{in}\,\partial m_{in}) \qquad ...(5.3)$$
$$= \text{shaft work} + \text{flow work}.$$

Now net work = $W_{net} = \partial W + (P_{out} v_{out}\,\partial m_{out} - P_{in} v_{in}\,\partial m_{in})$.

The first law as applied to flow system is

$$\partial Q = dE + \partial W_{net}$$
$$= (E_{\tau+\partial\tau} - E_\tau) + (e_{out}\,\partial m_{out} - e_{in}\,\partial m_{in}) + \partial W$$
$$+ (P_{out} v_{out}\,\partial m_{out} - P_{in} v_{in}\,\partial m_{in}) \qquad [P = \text{Pressure}, v = sp.\text{ vol}]$$

or
$$\partial Q + \partial m_{in}(e_{in} + P_{in} v_{in})$$
$$= (E_{\tau+\partial\tau} - E_\tau) + \partial m_{out}(e_{out} + P_{out} v_{out}) + \partial W \qquad ...(5.4)$$
$$[l = lk + lp + u]$$

Now
$$e + pv = \left(u + \frac{1}{2}\overline{V}^2 + gz\right) + pv$$
$$= (u + pv) + \frac{1}{2}\overline{V}^2 + gz$$
$$= h + \frac{1}{2}\overline{V}^2 + gz \qquad [\bar{v} = \text{velocity}]$$

Thus eqn. (5.4) becomes

$$\partial Q + \partial m_{in}\left(h + \frac{\overline{V}^2}{2} + gz\right)_{in} = (E_{\tau+\partial\tau} - E_\tau) + \partial W + \partial m_{out}\left(h + \frac{\overline{V}^2}{2} + gz\right)_{out}$$

Dividing by $\partial\tau$ and taking $\partial\tau \to 0$ we get

$$\dot{Q} + \dot{m}_{in}\left(h + \frac{\overline{V}^2}{2} + gz\right)_{in} = \left(\frac{dE}{d\tau}\right)_{CV} + \dot{m}_{out}\left(h + \frac{\overline{V}^2}{2} + gz\right)_{out} + \dot{W} \quad ...(5.5)$$

In case of a number of streams we change $\dot{m}_{in}$ in to $\Sigma\,\dot{m}_{in}$ and $\dot{m}_{out}$ to $\Sigma\,\dot{m}_{out}$ in the above equation. Equation (5.5) is the mathematical statement of the first law of thermodynamics for an OPEN SYSTEM.

The first law as applied to closed system is

$$đQ = dE + đW$$

or

$$\frac{đQ}{đ\tau} = \frac{dE}{d\tau} + \frac{đW}{đ\tau}$$

or

$$\dot{Q} = \frac{dE}{d\tau} + \dot{W} \quad ...(5.6)$$

In closed system $\quad m$ = constant

$\therefore \quad \dot{m}_{in} = \dot{m}_{out} = 0$, thus equation (5.5) reduces to equation (5.6).

5.4 STEADY-STATE STEADY FLOW ENERGY EQUATION

Steady flow means that the rates of flow of mass and energy across the control surface is constant.

Steady-state steady flow (SSSF) refers to all mass flow rates and all thermodynamic states remain steady and do not vary with time.

Thus

$$\left(\frac{dm}{d\tau}\right)_{CV} = 0, \quad \left(\frac{dE}{d\tau}\right)_{CV} = 0$$

So equation (5.2) becomes

$$\Sigma\,\dot{m}_{out} = \Sigma\,\dot{m}_{in} = \dot{m}\,(\text{say}) \quad ...(5.7)$$

and from equation (5.5) we get

$$\dot{Q} + \dot{m}\left(h + \frac{\overline{V}^2}{2} + gz\right)_{in} = \dot{m}\left(h + \frac{\overline{V}^2}{2} + gz\right)_{out} + \dot{W} \quad ...(5.8)$$

Let $\quad q = \dfrac{\dot{Q}}{\dot{m}}$ and $w = \dfrac{\dot{W}}{\dot{m}}$ we have

$$q + \left(h + \frac{\overline{V}^2}{2} + gz\right)_{in} = \left(h + \frac{\overline{V}^2}{2} + gz\right)_{out} + w \quad ...(5.9)$$

Unit of $$q = \frac{\dot{Q}}{\dot{m}} = \frac{KJ/S}{kg/S} = KJ/kg$$

Unit of $$w = \frac{\dot{W}}{\dot{m}} = \frac{KJ/S}{kg/S} = KJ/kg.$$

Thus q and w can be thought of as the heat transfer and flow work per unit mass flowing into and out of the control volume for this particular SSSF process.

Equation (5.8) is known as steady-state steady flow energy equation.

Equation (5.9) can be written in differential form as

$$đq = dh + \bar{V}d\bar{V} + gdz + đw. \quad \text{...(5.10)}$$

5.5 EXAMPLES OF STEADY FLOW PROCESSES

5.5.1 Nozzle and Diffuser

A nozzle is a device for increasing velocity or K.E. of a fluid at the expense of its pressure drop. A diffuser is a device for increasing the pressure of a fluid at the expense of its K.E.

Fig. 5.3 shows a convergent-divergent adiabatic nozzle using equation (5.9) between sections (1) and (2) we have

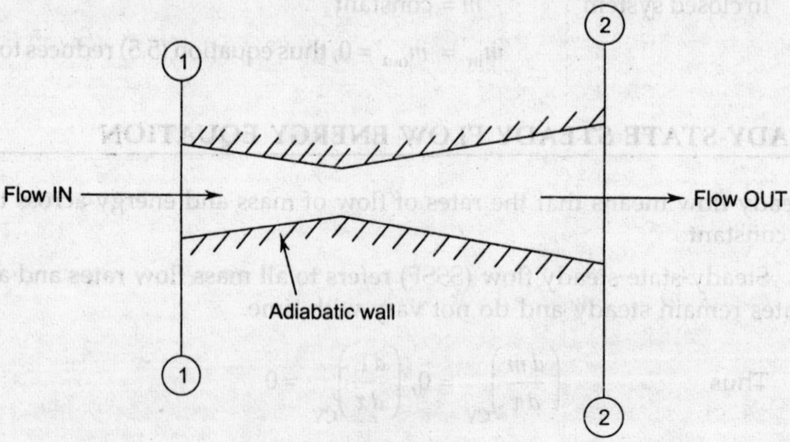

Fig. 5.3

$$q + \left(h + \frac{\bar{V}^2}{2} + gz\right)_{in} = \left(h + \frac{\bar{V}^2}{2} + gz\right)_{out} + w$$

Here $q = 0, w = 0$. Let us use subscript (1) for IN and (2) for OUT

$$\therefore \quad h_1 + \frac{\bar{V}_1^2}{2} + gz_1 = h_2 + \frac{\bar{V}_2^2}{2} + gz_2$$

$$z_1 = z_2$$

$$\therefore \quad h_1 + \frac{\bar{V}_1^2}{2} = h_2 + \frac{\bar{V}_2^2}{2}$$

or,
$$\overline{V}_2 = \sqrt{2(h_1 - h_2) + \overline{V}_1^2}\,.$$

The SSSF energy equation for an adiabatic diffuser is

$$h_2 - h_1 = \frac{\overline{V}_1^2 - \overline{V}_2^2}{2} \qquad ...(5.11)$$

i.e., there is rise in enthalpy and pressure at the expense of K.E. (kinetic energy decreases).

Application of Nozzle

(*i*) Steam nozzles used at inlet to turbines.
(*ii*) Rocket nozzles used in rocket engine, missiles, space vehicles.
(*iii*) Injector nozzles used in combustion chamber.

Application of Diffuser

(*i*) Used at the inlet section of an aircraft engine.
(*ii*) Used in centrifugal compressors.

5.5.2 Throttling Process

It is a process of reducing pressure. Fig. 5.4 shows a partially opened valve. The passage of fluid flow is restricted.

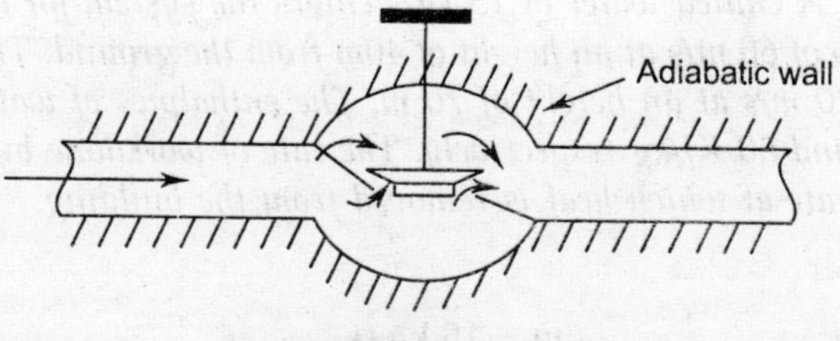

Fig. 5.4

Here $q = 0, w = 0$, and $\overline{V}_1 = \overline{V}_2$ thus $h_1 = h_2$.

Thus the throttling expansion process is an isenthalpic expansion process.

5.5.3 Steady Flow Through Heat Exchanger

In heat exchanger there are two fluids. One is a hot fluid (say A) and the other is a cold fluid (say B). Heat is transferred from hot fluid to the cold fluid. Thus heat lost by the hot fluid is equal to the heat gained by the cold fluid provided there is no loss of heat between the system and surroundings.

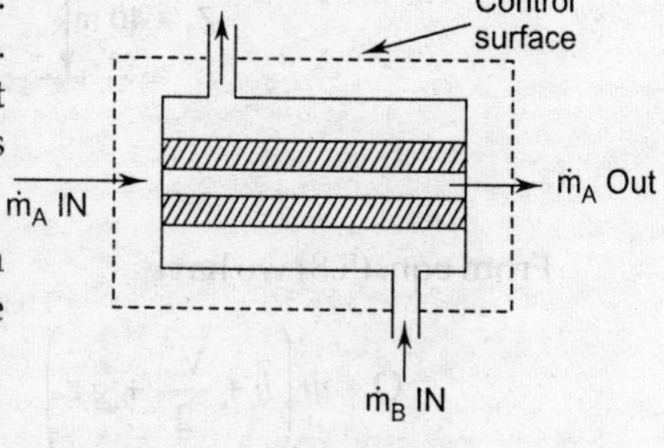

Fig. 5.5

Considering no heat and work transfer from the control surface and K.E. & P.E. change negligible. We can write down

$$(\dot{m}_A\, h_A)_{in} + (\dot{m}_B\, h_B)_{in}$$

$$= (\dot{m}_A h_A)_{out} + (\dot{m}_B h_B)_{out}$$

or $$\dot{m}_A \left(h_{A_{IN}} - h_{A_{out}}\right) = \dot{m}_B \left(h_{B_{out}} - h_{B_{IN}}\right)$$

$$= -\dot{Q}_A = \dot{Q}_B \quad ...(5.12)$$

Condenser of a steam power plant is the example of the heat exchanger.

5.5.4 Turbines and Compressors

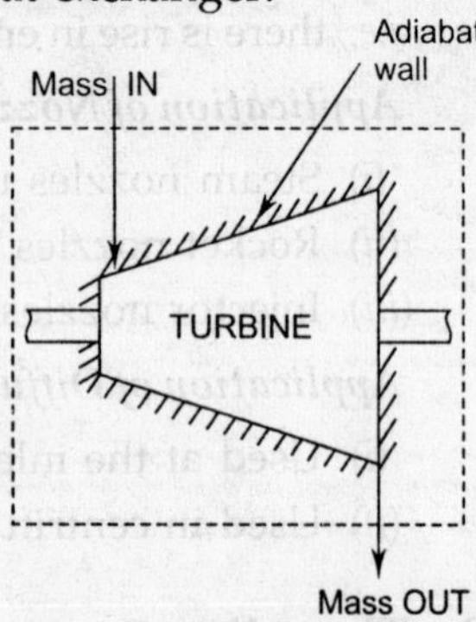

Fig. 5:6

We get positive work as an output for turbine.

We give work as input to the compressor.

Fig. 5.6 shows a turbine which is insulated.

Here $q = 0,\ \overline{V_1} = \overline{V_2},\ z_1 = z_2$. Thus

$$h_1 = h_2 + w$$

or $$w = (h_1 - h_2)$$

SOLVED EXAMPLES

Example 5.1. *A chilled water of 15 kg/s enters the system for air conditioning a tall building with a velocity of 60 m/s at an height of 40m from the ground. The water leaves the system with a velocity of 20 m/s at an height of 70 m. The enthalpies of water entering in and leaving out are 30 KJ/kg and 50 KJ/kg respectively. The rate of workdone by a pump in the line is 40 KW. Find out the rate at which heat is removed from the building.*

Solution.

Given $\dot{m} = 15$ kg/s

W = – 40 KW

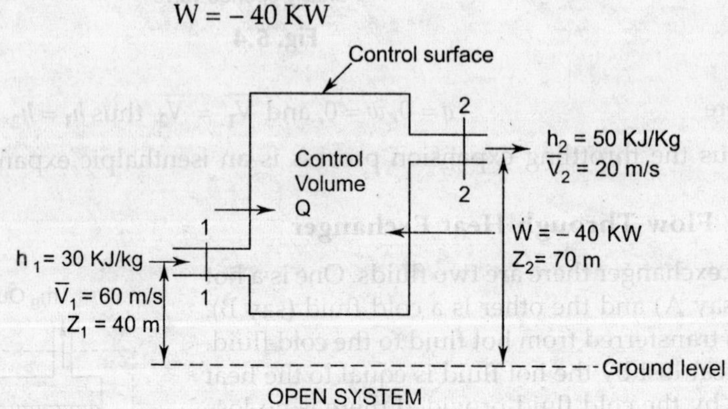

Fig. 5.7

From eqn. (5.8) we have

$$\dot{Q} + \dot{m}\left(h + \frac{\overline{V}^2}{2} + gz\right)_{in} = \dot{m}\left(h + \frac{\overline{V}^2}{2} + gz\right)_{out} + \dot{W}$$

or,
$$\dot{Q} = \dot{m}\,(h_2 - h_1) + \dot{m}\left(\frac{\overline{V}_2^2 - \overline{V}_1^2}{2}\right) + \dot{m}\,g(z_2 - z_1) + \dot{W}$$

$$= 15(50-30)+15\left(\frac{20^2-60^2}{2}\right)\frac{1}{1000}+\frac{15\times 9.81(70-40)}{1000}-40$$

$$= 240.41 \text{ KW} = \text{heat removal rate.} \quad \textbf{Ans.}$$

Example 5.2. *A steam turbine receives steam at the rate of 0.42 kg/s. The inlet and outlet conditions of steam are as follows.*

	Inlet	*Outlet*
Pressure	*1.2 MPa*	*20 KPa*
Temperature	*188ºC*	–
Enthalpy	*2785 KJ/kg*	*2512 KJ/kg*
Velocity	*33.3 m/s*	*100 m/s*
Elevation	*3 m*	*0 m*

If the heat lost to the surroundings is 0.29 KJ/s, find out the power output of the turbine under steady flow conditions.

Solution. Given $\dot{Q} = -0.29$ KJ/s; $\dot{m} = 0.42$ kg/s, $h_1 = 2785$ KJ/kg, $V_1 = 33.3$ m/s, $z_1 = 3m$, $h_2 = 2512$ KJ/kg, $\overline{V}_2 = 100$ m/s, $z_2 = 0$ m

we have

$$\dot{Q} + \dot{m}\left(h_1 + \frac{\overline{V}_1^2}{2} + g\,z_1\right) = \dot{m}\left(h_2 + \frac{\overline{V}_2^2}{2} + g\,z_2\right) + \dot{W}$$

$$-0.29+0.42\left(2785+\frac{33.3^2}{2000}+\frac{9.81\times 3}{1000}\right) = 0.42\left(2512+\frac{100^2}{2000}+\frac{9.81\times 0}{1000}\right)+\dot{W}$$

or
$$\dot{W} = 112.51 \text{ KW} = \text{Power output from the turbine.} \quad \textbf{Ans.}$$

Example 5.3. *An adiabatic horizontal nozzle receives steam at velocity of 60 m/s and enthalpy of 3000 KJ/kg. It discharges it at an enthalpy 2762 KJ/kg, (a) find the velocity at exist from the nozzle (b) if the inlet area is 0.1 m² and the specific volume at inlet is 0.187 m³/kg, find the mass flow rate (c) if the specific volume at the nozzle exit is 0.498 m³/kg, find the exit area of the nozzle.*

Solution. For adiabatic ($\dot{Q} = 0$), horizontal ($z_1 = z_2$) nozzle; we have

$$h_2 - h_1 = \frac{\overline{V}_1^2 - \overline{V}_2^2}{2}$$

or
$$2762 - 3000 = \frac{60^2 - \overline{V}_2^2}{2000}$$

or
$$\overline{V}_2 = 692.5315\, m/s. \quad \textbf{Ans. (a)}$$

$$\dot{m} = \frac{A_1\,\overline{V_1}}{v_1} = \frac{0.1\times 60}{0.187} = 32.08\, kg/s. \quad \textbf{Ans. (b)}$$

also $$\dot{m} = \frac{A_2 \overline{V_2}}{v_2}$$

or $$32.08 = \frac{A_2 \times 692.5315}{0.498}, A_2 = 0.023\, m^2. \quad \textbf{Ans. (c)}$$

Example 5.4. *In an oil cooler, oil flows steadily through a bundle of metal tubes submerged in a steady stream of cooling water. Under steady flow conditions, the oil enters at 90°C and leaves at 30°C, while water enters at 25°C and leaves at 70°C. The enthalpy of oil at t°C is given by*

$$h = 1.68\, t + 10.5 \times 10^{-4}\, t^2 \text{ KJ/kg.}$$

What is the cooling water flow required for cooling 2.78 kg/s of oil ?

Solution.

$$h = 1.68\, t + 10.5 \times 10^{-4}\, t^2$$

$$h_{A\,in} = 1.68 \times 90 + 10.5 \times 10^{-4} \times 90^2 = 159.7 \text{ KJ}/kg$$

$$h_{A\,out} = 1.68 \times 30 + 10.5 \times 10^{-4} \times 30^2 = 51.345 \text{ KJ}/kg$$

From equation (5.12) we have

$$\dot{m}_A\ (h_{A\,in} - h_{A\,out}) = \dot{m}_B\ (h_{B\,out} - h_{B\,in})$$

or $$2.78\,(159.7 - 51.345) = \dot{m}_B \times C_W\,(t_{B\,out} - t_{B\,in})$$

[A = hot liquid = oil, B = Water, *i.e.*, cold liquid]

or $$2.78\ \ (159.7 - 51.345) = \dot{m}_B\ \times 4.1868\,(70 - 25)$$

or $$\dot{m}_B = 1.59 \text{ kg/s.} \quad \textbf{Ans.}$$

Chapter 6

Second Law of Thermodynamics

6.1 CONVERSION OF WORK INTO HEAT

Work can be converted into equal amount of heat. So methematically we can write $W \underset{\rightarrow}{=} Q$. Where → indicates the direction of energy conversion.

Fig. 6.1 shows that process 1 to 2 is done when W_{1-2} work is given as an input. The process 2 to 1 is obtained by heat transfer Q_{2-1}. It has been found that

$$W_{1-2} = Q_{2-1}.$$

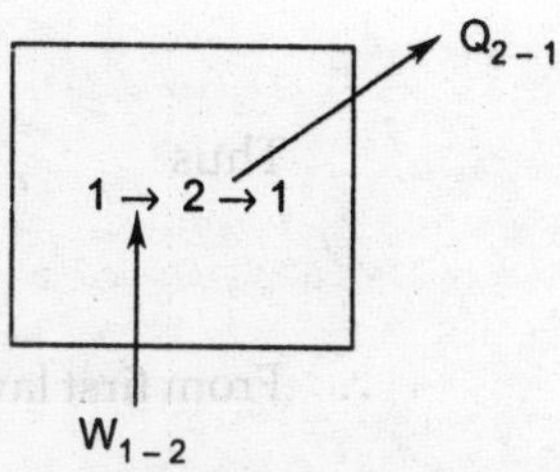

Fig. 6.1

6.2 CONVERSION OF HEAT INTO WORK

Spontaneous processes in nature occur only in one direction. Heat always flows from a body at a higher temperature to a body at a lower temperature. If we want to transfer equal heat from lower temperature body to higher temperature body as a spontaneous process, we cannot do it because some work is to be done. First law does not restrict on direction of conversion of heat to work. Heat cannot be converted to equal amount of work.

i.e., $Q \underset{\rightarrow}{\geq} W$. Some heat is to be rejected. The amount of heat rejected is Q – W.

Fig. 6.2. shows that process 1 to 2 is carried out by injecting heat Q_{1-2}. If the process 2 to 1 is obtained by doing some work on it *i.e.*, W_{2-1}, it is found that

$$Q_{1-2} > W_{2-1}.$$

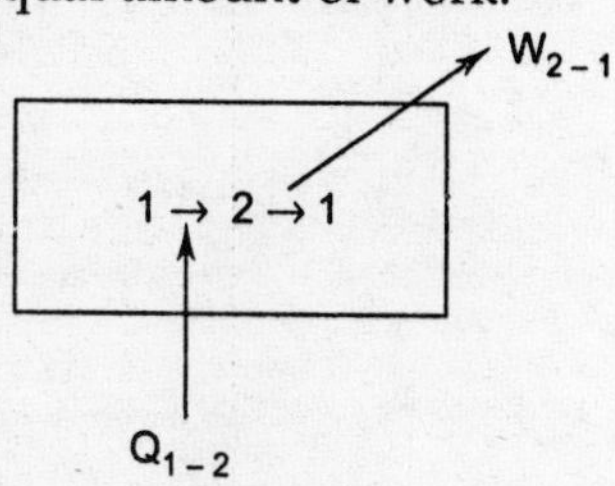

Fig. 6.2

6.3 KELVIN-PLANK STATEMENT OF SECOND LAW OF THERMODYNAMICS

6.3.1 Statement

Kelvin-Plank states, *"It is impossible to construct an engine working in a cycle whose sole effect is the conversion of all the heat energy supplied to it by a source into an equivalent amount of work"*.

6.3.2 Explanation

Heat Q_1 is given to the system, the system does some work W_E. To bring back to original position of the piston W_C work is done on the system as a result Q_2 amount of heat is rejected from the system.

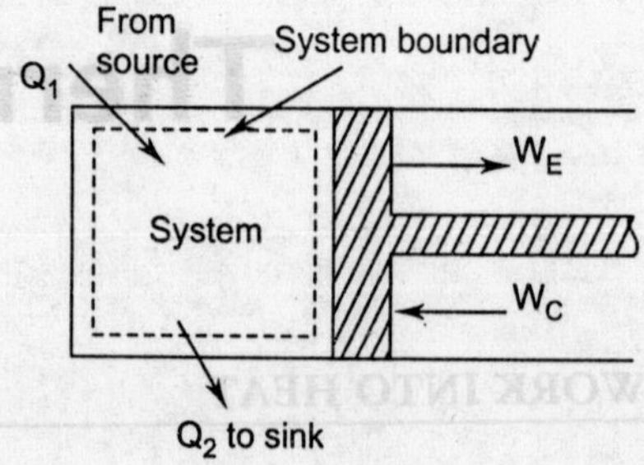

Fig. 6.3

Thus

$$Q_{net} = Q_1 - Q_2$$
$$W_{net} = W_E - W_C$$
$$\Sigma Q_{cycle} = \Sigma W_{cycle}$$

$\therefore$ From first law, cycle or $Q_{net} = W_{net}$

$$Q_1 - Q_2 = W_E - W_C.$$

Fig. 6.4 shows a heat engine cycle

Process 1 – 2 = Q_1 heat is added from furnace to boiler

Process 2 – 3 = W_T work is done by Turbine.

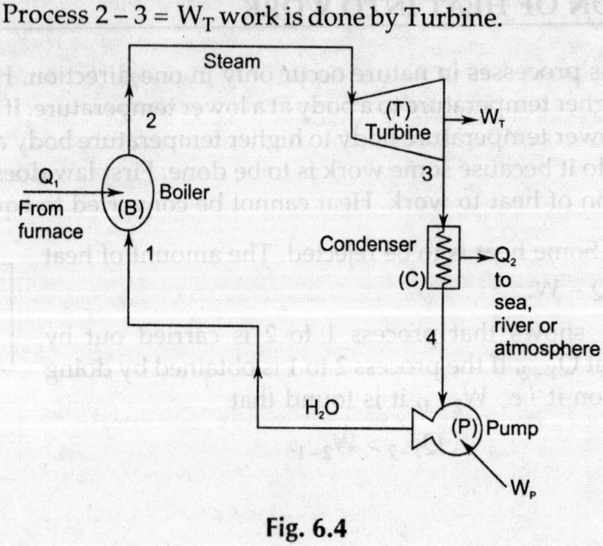

Fig. 6.4

Process 3 – 4 = Q_2 heat is rejected to sea, river or atmosphere.

Process 4 – 1 = W_P work is done on the pump.

Thus the cycle completes.

$$\sum_{cycle} Q = \sum_{cycle} W$$

$$Q_1 - Q_2 = W_T - W_P.$$

6.3.3 Heat Engine

Heat engine is a device which works in a complete cycle and converts heat to work. Thus $(W_T - W_P)$ (Ref Fig. 6.4) and Q_1 are of primary interest for heat engine. The efficiency of heat engine is defined as

$$\eta = \frac{W_{net}}{Q_1} = \frac{W_T - W_P}{Q_1} = \frac{Q_1 - Q_2}{Q_1} = 1 - \frac{Q_2}{Q_1}.$$

This efficiency is termed as thermal efficiency of a heat engine cycle.

6.3.4 Thermal and Mechanical Energy Reservoirs

A thermal energy reservoir (TER) is a body with a very large heat capacity to which, and from which, heat can be transferred without any change in its temperature (thermodynamic property). Higher temperature body is written as TER_H and lower temperature body is written as TER_L. If heat is transferred from it, then it is known as heat source or simply source. If heat is transferred to it, then it is considered as a heat sink or simply sink. Heat reservoirs are characterised by its temperature. A mechanical energy reservoir (MER) is a large body enclosed by an adiabatic impermeable wall capable of storing work as potential energy (such as wound spring) or K.E. (such as rotating flywheel). An MER performs quasi-statically.

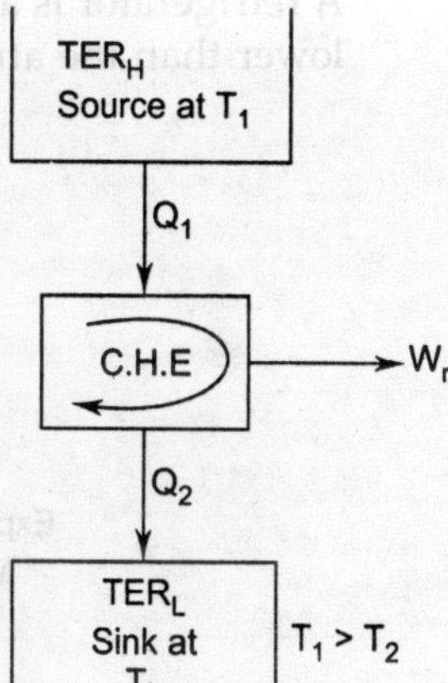

Fig. 6.5

Fig. 6.5 shows a cyclic heat engine (C.H.E). It takes Q_1 heat from the source and rejects Q_2 heat to the sink. It also performs work Wnet.

C.H.E. = Cyclic heat engine.

6.3.5 Perpetual Motion Machine of the Second Kind (PMM 2)

PMM 2 is a machine which takes heat from a single source and converts to equal amount of work as its sole effect. PMM 2 is impossible according to Kelvin-Plank's statement of the second law of thermodynamics.

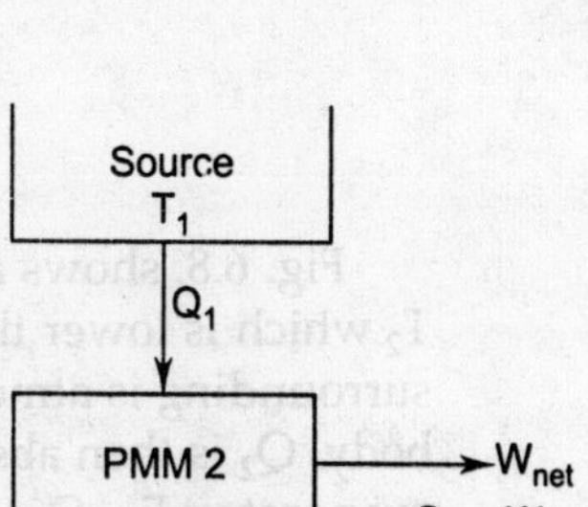

Fig. 6.6

6.4 CLAUSIUS' STATEMENT OF THE SECOND LAW

6.4.1 Statement

"It is impossible to construct a device which, operating in a cycle, will transfer heat from lower temperature body to a higher temperature body at its sole effect.

6.4.2 Explanation

Heat cannot flow of itself from a lower temperature body to a body at a higher temperature. Some work is to be done.

Fig. 6.7 shows that W is to be done on the system R to transfer Q_1 amount of heat to a body at T_1.

Here $Q_1 - Q_2 = W$.

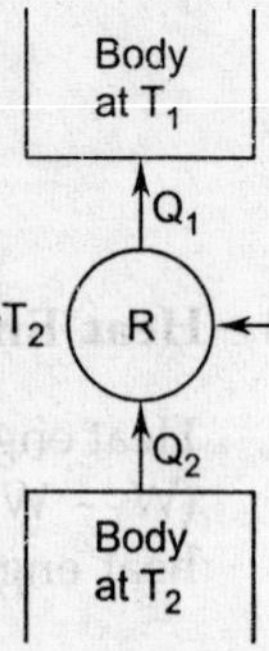

Fig. 6.7

6.5 REFRIGERATOR AND HEAT PUMP

6.5.1 Refrigerator

A refrigerator is a device which operating in a cycle, maintains a body at a temperature lower than the atmosphere or surroundings.

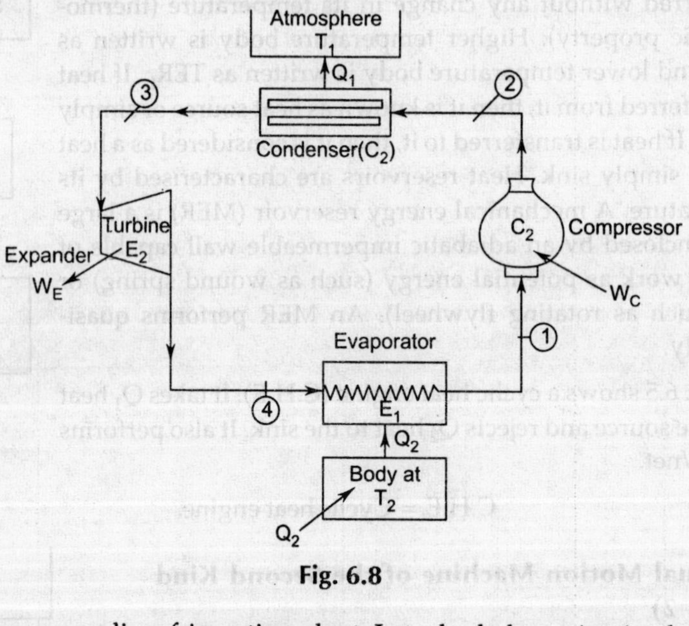

Fig. 6.8

Fig. 6.8. shows a cyclic refrigeration plant. Let a body be maintained at temperature T_2 which is lower than ambient temperature T_1. Q_2 is extracted from the body at T_2. As surrounding is atmosphere at temperature T_1 (> T_2), there will be heat leakage Q_2 to the body. Q_2 is then absorbed by the refrigerant (working fluid) and gets evaporated in the evaporator E_1. Q_2 supplies the latent heat of evaporation. The body thus maintains temperature T_2 lower than T_1 (process 4 – 1). The vapour is then compressed by C_1 which absorbs work W_C (process 1 – 2). It is condensed in C_2 rejecting heat Q_1 to atmosphere (process 2 – 3). The condensate then expands adiabatically through turbine E_2 producing work W_E, also the temperature drops to a value lower than T_2 (process 3 – 4) such that Q_2 flows from the body to evaporate again the refrigerant. $E_1 - C_1 - C_2 - E_2$ cyclic flow process is termed refrigeration and the device is refrigerator.

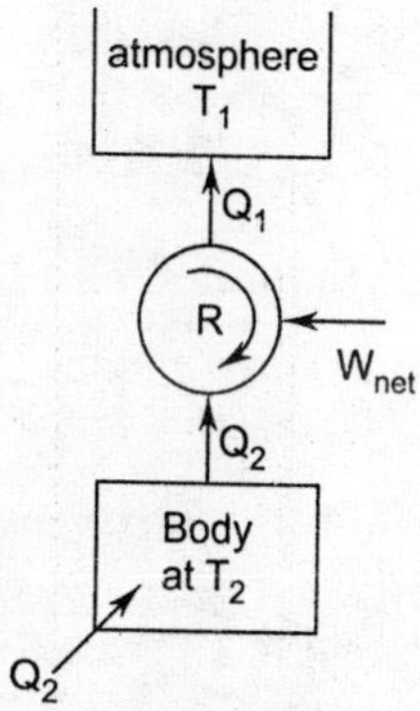

Fig. 6.9

Coefficient of performance (COP) is defined as

$$[COP]_{Ref} = \frac{Q_2}{W_{net}} = \frac{Q_2}{Q_1 - Q_2} \quad \text{...(6.1)}$$

where $$W_{net} = W_C - W_E = Q_1 - Q_2.$$

6.5.2 Heat Pump

Heat pump is a device which operates in a cycle and maintains a body at a temperature higher than the temperature of the atmosphere (surroundings).

$$[COP]_{HP} = \frac{Q_1}{W_{net}} = \frac{Q_1}{Q_1 - Q_2} \quad \text{...(6.2)}$$

Subtracting (6.1) from (6.2) we get

$$[COP]_{HP} - [COP]_{Ref} = 1 \quad \text{...(6.3)}$$

Fig. 6.10 shows cyclic heat pump (C.H.P). The C.H.P continuously supply heat of amount Q_1 to the body which maintains temperature at T_1. W_{net} work is done on the C.H.P.

$$W_{net} = W_C - W_E$$

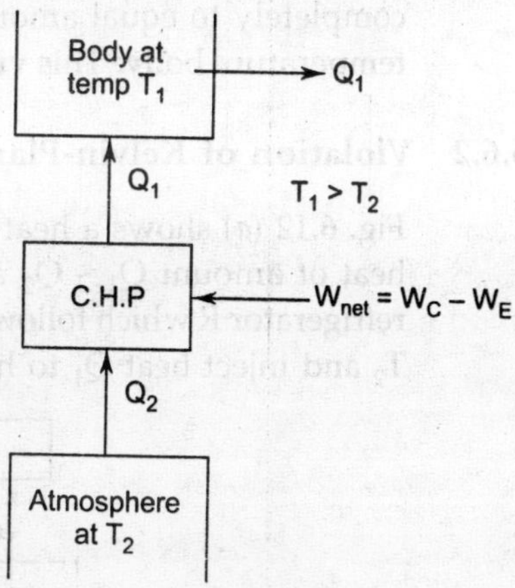

Fig. 6.10

6.6 EQUIVALENCE OF KELVIN-PLANK AND CLAUSIUS STATEMENTS

Though the two statements appear to be different, but they are equivalent, they mean the same thing. This is verified as follows.

6.6.1 Violation of Clausius Statement Leads to Violation of Kelvin-Plank Statement

Let us take a refrigerator 'R' which violates Clausius statement requiring no work to transfer heat Q_2 from low temperature T_2 to high temperature T_1.

Now we take a heat engine (E) which follows Kelvin-Plank statement *i.e.*, it draws heat Q_1 from T_1 and rejects heat Q_2 to T_2. Thereby it produces work W which is equal to $Q_1 - Q_2$ (Ref Fig. 6.11 (*a*)).

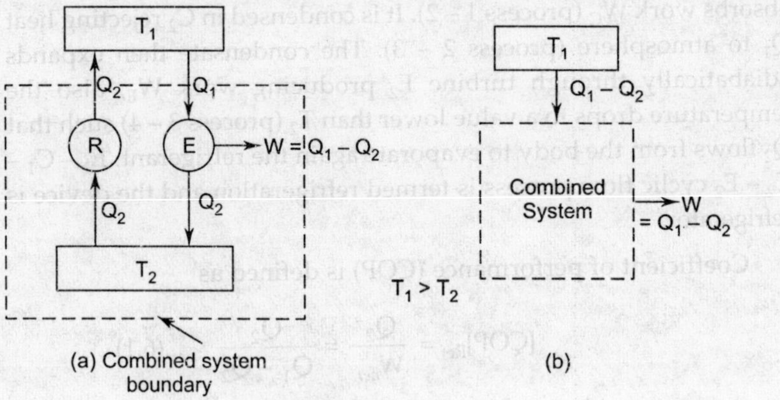

Fig. 6.11

Now let us couple the two engines and consider a combined system as shown in Fig. 6.11 (*b*). The combined system draws heat from T_1 which is equal to $Q_1 - Q_2$ and converts completely to equal amount of work $W = Q_1 - Q_2$ without rejecting any heat to lower temperature body. This violates Kelvin-Plank statement.

6.6.2 Violation of Kelvin-Plank Statement Leads to Violation of Clausius Statement

Fig. 6.12 (*a*) shows a heat engine E which violates Kelvin-Plank statement *i.e.*, it draws heat of amount $Q_1 - Q_2$ and converts completely to work $W = Q_1 - Q_2$. Let us take a refrigerator R which follows Clausius statement *i.e.*, it draws heat Q_2 from low temperature T_2 and inject heat Q_1 to high temperature T_1 and it requires work $W = Q_1 - Q_2$.

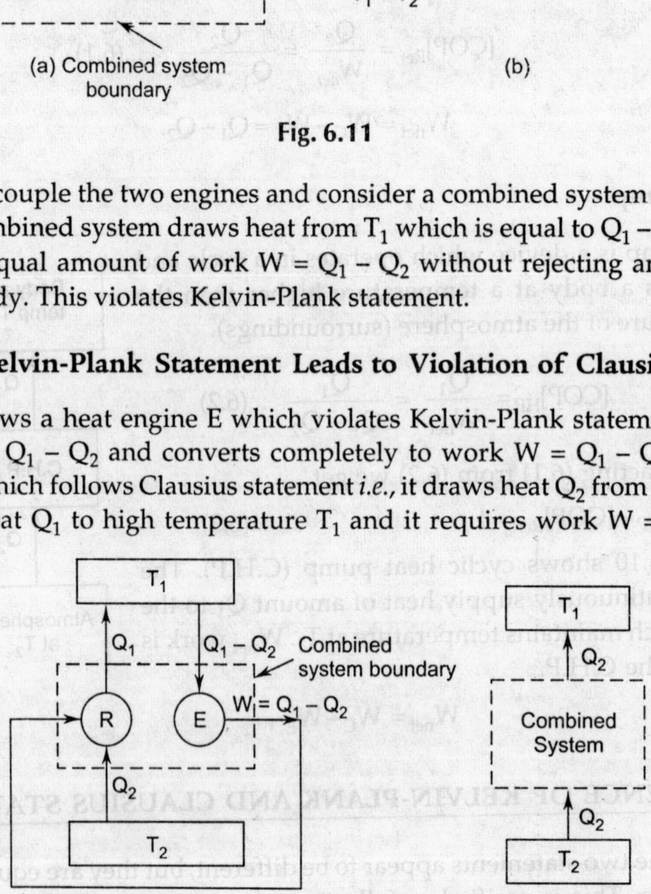

Fig. 6.12

If we couple R with E, the work from E is directly given to R. The combined system is shown in Fig. 6.12 (*b*). Fig. 6.12 (*b*) shows that the combined system draws heat of amount Q_2 from low temperature T_2 and inject same amount Q_2 to higher temperature T_1 without any workdone on it. Thus this violates Clausius statement.

SOLVED EXAMPLES

Example 6.1. *An engine inventor claims to have developed an engine that takes in 140 MJ at a temperature of 405 K and rejects 45 MJ at a temperature of 205 K while producing 20 KWh of mechanical work. Find if the claim of inventor is true or false.*

Solution. $Q_1 = 140$ MJ, $Q_2 = 45$ MJ, $W = 20$ KWh

Now $W = 20 \text{ KWh} = 20 \times 3600 \text{ KJ} = 72 \text{ MJ}.$

Now $W + Q_2 = (72 + 45) \text{ MJ} = 117 \text{ MJ}$

and $Q_1 = 140 \text{ MJ}$

Thus $Q_1 \neq W + Q_2 \rightarrow$ Violating 1st law of thermodynamics (energy not conserved)

$$(\eta_{Th})_{max} = 1 - \frac{T_2}{T_1} = 1 - \frac{205}{405} = 0.49$$

$$(\eta_{Th})_{claimed} = \frac{W}{Q_1} = \frac{72}{140} = 0.51$$

Thus $(\eta_{Th})_{claimed} > (\eta_{Th})_{max} \rightarrow$ It violates the 2nd law also.

Thus the claim is false.

Example 6.2. *If a refrigerator is sued for heating purposes in winter so that the atmosphere becomes the cold body and the room to be heated becomes the hot body, how much heat would be available for heating for every 1.5 KW input to the driving motor? The COP of the refrigerator is 6, and the electromechanical efficiency of the motor is 80%.*

Solution.

Here motor efficiency = 0.8

$$\frac{\text{Motor output}}{\text{input}} = 0.8$$

or $$\frac{\text{Motor output}}{1.5} = 0.8$$

$$\text{Motor output} = 0.8 \times 1.5 \text{ KW}$$
$$= W \text{ to Refrigerator}$$

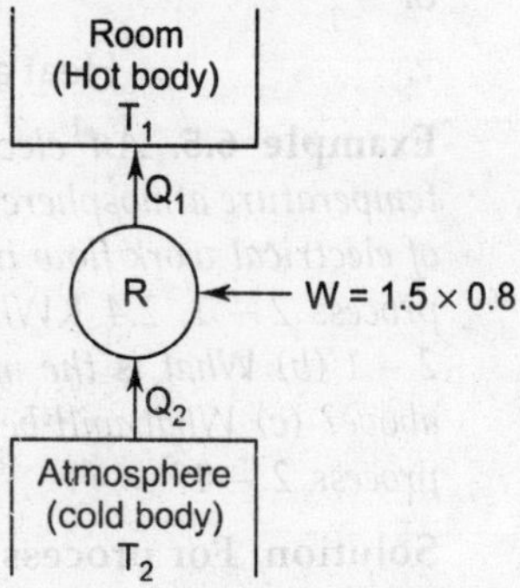

Fig. 6.13

$$COP_{ref} = \frac{Q_2}{W}$$

or $$6 = \frac{Q_2}{0.8 \times 1.5} \text{ or } Q_2 = 7.2 \text{ KW}$$

$\therefore$ $$Q_1 = W + Q_2 = 0.8 \times 1.5 + 7.2$$
$$= 8.4 \text{ KW}$$

Heat available = 8.4 KW. **Ans.**

Example 6.3. *A refrigerator with COP of 5 is driven by an engine of 30% thermal efficiency. What is the heat input into the engine for each MJ removed from the cold body by the refrigerator?*

Solution. $COP_{ref} = \frac{Q_2}{W}$

or $5 = \frac{1}{W}$ or $W = \frac{1}{5} = 0.2$ MJ = output of the engine

For engine $\frac{\text{output}}{\text{input}} = 0.3$

or $\frac{0.2}{\text{input of engine}} = 0.3$

or Input of engine $= \frac{0.2}{0.3}$ MJ = 666.67 KJ. **Ans.**

Example 6.4. *If the system (refer to Example 6.3) is used as a heat pump, how many MJ of heat would be available for heating for each MJ of heat input to the engine?*

Solution. $\frac{\text{Output of engine}}{\text{Input of engine}} = 0.3$

or output of engine = 0.3 × input of engine

$= 0.3 \times 1 = 0.3 \text{ MJ} = W$

$COP_{ref} = 5 = \frac{Q_2}{W} = \frac{Q_2}{0.3}$

or $Q_2 = 5 \times 0.3 = 1.5$ MJ

∴ Heat available $= Q_2 + W = 1.5 + 0.3 = 1.8$ MJ. **Ans.**

Example 6.5. *An electric storage battery which can exchange heat only with a constant temperature atmosphere goes through a complete cycle of two processes. In process 1 – 2, 2.8 KWh of electrical work flow into the battery while 732 KJ of heat flow out to the atmosphere. During process 2 – 1, 2.4 KWh of work flow out of the battery. (a) Find the heat transfer in process 2 – 1 (b) What is the maximum possible work of 2 – 1 process if 1 – 2 process has occured as above? (c) What will be the heat transfer in the condition of maximum possible work output in process 2 – 1?*

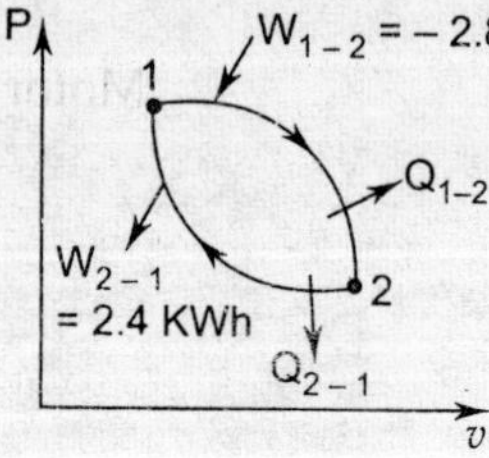

Fig. 6.14

Solution. For process 1 – 2

$Q_{1-2} = U_2 - U_1 + W_{1-2}$

or $-732 = U_2 - U_1 - 2.8 \times 3600$

or $U_2 - U_1 = 9348$ KJ

For process 2 – 1

$Q_{2-1} = U_1 - U_2 + W_{2-1}$

$= -9348 + 2.4 \times 3600$

$= -708$ KJ **Ans. (a)**

(Negative sign shows that heat flow out of the system)

$Q_{2-1} = U_1 - U_2 + W_{2-1}$

For maximum work $Q_{2-1} = 0$

or $$(W_{2-1})_{max} = U_2 - U_1 = 9348 \text{ KJ}. \quad \textbf{Ans. (b)}$$

$$Q_{2-1} = 0 \text{ for maximum work}. \quad \textbf{Ans. (c)}$$

Example 6.6. *The foods kept inside the household refrigerator is maintained at temperature of 2°C. A man opens the door 20 times in a day of April introducing 420 KJ heat each time to the foods. The actual operating COP of the refrigerator is only 15% of the ideal COP. The cost of work is Rs. 2.50 per KWh. What is the bill for this refrigerator in the month end of April. The ambient temperature is 30°C.*

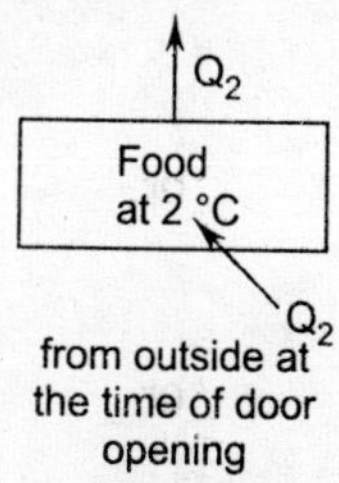

Fig. 6.15

Solution.
$$T_2 = 2 + 273 = 275 \text{ K}$$

$$T_1 = 30 + 273 = 303 \text{ K}.$$

$$(COP)_{ideal} = \frac{Q_2}{Q_1 - Q_2} = \frac{T_2}{T_1 - T_2} = \frac{275}{28} = 9.82$$

∴ $$(COP)_{actual} = 0.15 \times 9.82 = 1.473$$

Given $$Q_2 = 420 \times 20 \times 30 \text{ KJ per month. [April = 30 days]}$$

∴ $$\frac{Q_2}{W} = 1.473$$

or $$W \text{ (per month)} = \frac{Q_2}{1.473} = \frac{420 \times 20 \times 30}{1.473}$$

$$= 171079.42 \text{ KJ}$$

∴ $$\text{Bill for April} = \frac{171079.42}{3600} \times 2.5$$

$$= \text{Rs. } 118.805 = \text{Rs. } 118.80. \quad \textbf{Ans.}$$

Example 6.7. *A heat engine working on Carnot cycle converts one-fifth of the heat input into work. When the temperature of the sink is reduced by 80°C, the efficiency gets doubled. Calculate the temperature of source and sink.*

Solution. The Carnot heat engine is a reversible engine and its efficiency is given by

$$\eta = 1 - \frac{T_2}{T_1}$$

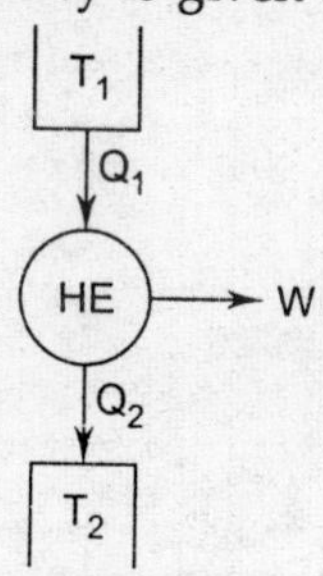

Fig. 6.16

In the first case $$\frac{W}{Q_1} = \frac{1}{5} = \frac{\text{Net work}}{\text{Heat supplied}}$$

$$\frac{Q_1 - Q_2}{Q_1} = \frac{1}{5}$$

or $$1 - \frac{Q_2}{Q_1} = \frac{1}{5}$$

or $$1 - \frac{T_2}{T_1} = \frac{1}{5}$$

or $$\frac{T_2}{T_1} = 1 - \frac{1}{5} = \frac{4}{5} \qquad ...(1)$$

In the 2nd case

$$1 - \frac{(T_2 - 80)}{T_1} = \frac{2}{5}$$

or $$\frac{T_2 - 80}{T_1} = 1 - \frac{2}{5} = \frac{3}{5}$$

or $$\frac{T_2}{T_1} - \frac{80}{T_1} = \frac{3}{5}$$

or $$\frac{4}{5} - \frac{80}{T_1} = \frac{3}{5}, \text{ or } \frac{80}{T_1} = \frac{4}{5} - \frac{3}{5} = \frac{1}{5}$$

or $$T_1 = 400 \text{ K}$$

$$T_2 = \frac{4}{5} \times 400 = 320 \text{ K.} \quad \textbf{Ans.}$$

Chapter 7

Carnot Cycle

7.1 INTRODUCTION OF CARNOT CYCLE

Carnot cycle was introduced by Nicolas Leonard Sadi Carnot (1796–1834), a French Engineer, in 1824. It consists of four reversible processes, two frictionless isothermal and two frictionless adiabatics. He used piston-cylinder mechanism and made the following assumptions.

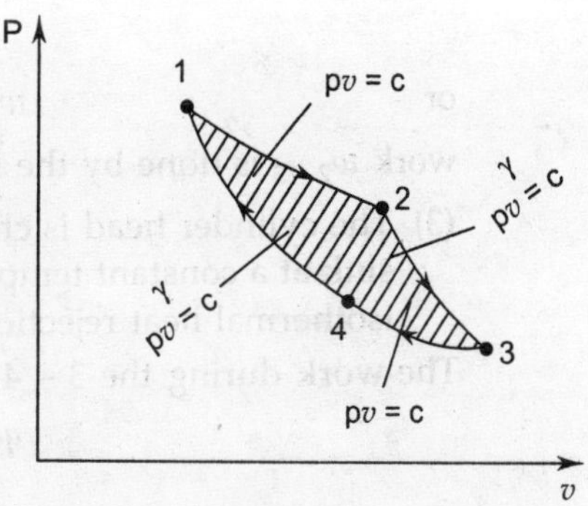

Fig. 7.1

- The piston moving in cylinder does not encounter any friction during motion.
- The walls of piston and cylinder are perfect insulators of heat.
- The head of the cylinder is so arranged that it acts as perfect conductor and also as a perfect insulator (Interchangeable cylinder head).
- The heat transfer to and from source & sink does not change the temperature of source and sink.

7.2 WORKING OF THE CYCLE

The working of Carnot cycle is shown in Figs. 7.1 & 7.2 on P-v diagram. The sequence of operations is described below.

(1) State 1 is having initial volume and initial pressure of the 1 kg mass system in the cylinder.

Now the cylinder head is a perfect conductor of heat and q_1 quantity of heat at T_1 (source temperature) is supplied to system of the engine cylinder till the state 2 is reached. The volume of the system increases and isothermal expansion process 1 – 2 takes place. At point 2, the supply of heat is cut off.

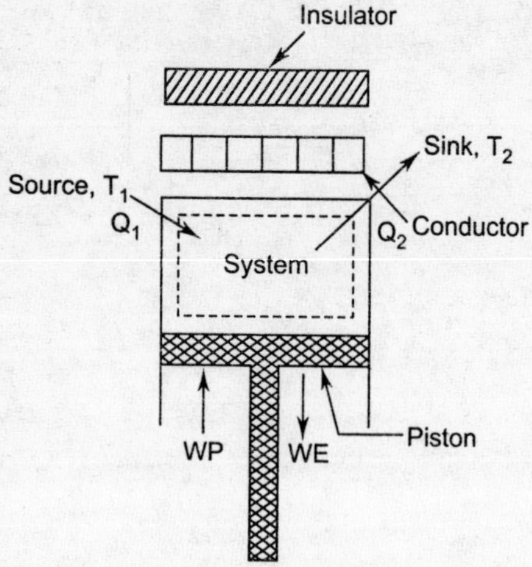

Fig. 7.2

For process 1 – 2 we apply 1st law,

$$q_{1-2} = u_2 - u_1 + w_{1-2}$$
$$[u_2 - u_1 = 0 \text{ for isothermal process}]$$
$$= 0 + w_{1-2}$$

or simply $q_1 = P_1 v_1 \ln \dfrac{v_2}{v_1} = RT_1 \ln \dfrac{v_2}{v_1}$ ...(7.1)

w_{1-2} is the workdone by the system it is positive $(v_2 > v_1)$.

(2) At the point 2, cylinder head is changed to insulator. During the process 2 – 3, the heat source is removed, the system expands reversibly and adiabatically till its temperature falls to T_2.

For process 2 – 3

$$q_{2-3} = u_3 - u_2 + w_{2-3}$$
$$0 = u_3 - u_2 + w_{2-3}$$

or
$$w_{2-3} = (u_2 - u_3)$$

work w_{2-3} is done by the system at the expense or decrease in internal energy.

(3) The cylinder head is changed to conductor of heat. q_2 amount of heat is rejected to sink at a constant temperature T_2. The process 3 – 4 is now followed and it is again isothermal heat rejection to sink.

The work during the 3 – 4 process is given by

$$q_{3-4} = u_4 - u_3 + w_{3-4}$$

or
$$q_2 = 0 + w_{3-4} = P_3 v_3 \ln \frac{v_4}{v_3}$$

or
$$-q_2 = P_3 v_3 \ln \frac{v_3}{v_4} = RT_2 \ln \frac{v_3}{v_4}.$$

This w_{3-4} is done on the system. ...(7.2)

(4) The cylinder head is again changed to insulator and adiabatic compression is done to bring back to state 1. The cycle is completed.

During process 4 – 1 the workdone on the system is given by

$$q_{4-1} = u_1 - u_4 + w_{4-1}$$
$$0 = u_1 - u_4 + w_{4-1}$$

or
$$w_{4-1} = u_4 - u_1.$$

[**Note.** Lower case letters are used for 1 kg of the mass of the system.]

7.3 EFFICIENCY OF CARNOT CYCLE

$$\eta_{Carnot} = \frac{W_{net}}{q_1} = \frac{q_1 - q_2}{q_1} = 1 - \frac{q_2}{q_1}$$

$$= 1 - \frac{RT_2 \ln \frac{v_3}{v_4}}{RT_1 \ln \frac{v_2}{v_1}} \quad ...(7.3)$$

Applying adiabatic law to process 2 – 3 and 4 – 1 we get

$$P_2 v_2^{\gamma} = P_3 v_3^{\gamma}$$

$$P_2 v_2 v_2^{\gamma-1} = P_3 v_3 v_3^{\gamma-1}$$

$$RT_1 v_2^{\gamma-1} = RT_2 v_3^{\gamma-1}$$

or

$$\frac{T_1}{T_2} = \left(\frac{v_3}{v_2}\right)^{\gamma-1} \quad ...(7.4)$$

Similarly for process 4 – 1

$$\frac{T_1}{T_2} = \left(\frac{v_4}{v_1}\right)^{\gamma-1} \quad ...(7.5)$$

Equating equations (7.4) and (7.5) we get

$$\frac{v_3}{v_2} = \frac{v_4}{v_1}$$

or

$$\frac{v_2}{v_1} = \frac{v_3}{v_4} \quad ...(7.6)$$

Thus equation (7.3) becomes

$$\boxed{\eta_{Carnot} = 1 - \frac{T_2}{T_1}}$$

where T_1 = source temperature and T_2 = Sink temperature

Thus Carnot efficiency depends on source and sink temperatures only and does not depend on the working fluid of the system. To increase the efficiency of Carnot cycle, T_1 is to be increased because T_2 cannot be decreased as generally atmosphere is taken to be sink.

Drawbacks of Carnot Cycle

Following are the drawbacks of Carnot Cycle :

- Frictionless motion of the piston through cylinder is impossible.
- An interchangable cylinder head makes engine impractical.
- Isothermal expansion is a slow process and adiabatic expansion should be very fast. This is impractical in a single stroke of the piston.

7.4 CARNOT'S THEOREM

Theorem 1. *"No heat engine working in a cyclic process between two fixed temperatures can be more efficient than a reversible heat engine working between the same two fixed temperatures."*

Proof: Refer Fig. 7.3 (*a*)

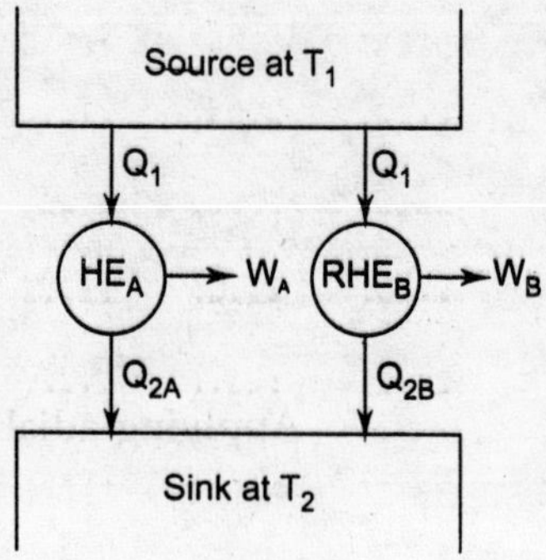

Fig. 7.3(*a*)

Let HE_A = Any heat engine

RHE_B = Reversible heat engine and both the engine draw Q_1 amount a heat. Suppose the statement is not true, so, $\eta_A > \eta_B$

or $$\frac{W_A}{Q_1} > \frac{W_B}{Q_1}$$

or $$W_A > W_B.$$

Now we reverse the RHE_B as it is reversible. So all directions of flow will be reversed.

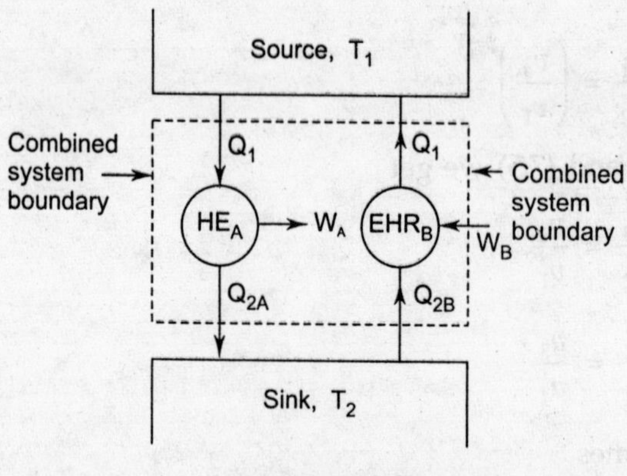

Fig. 7.3(*b*)

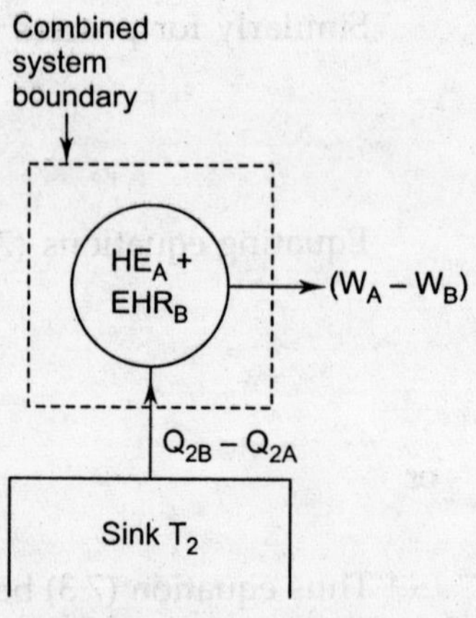

Fig. 7.3(*c*)

If we combine HE_A & EHR_B, we get work $W_A - W_B$ out of exchange heat $(Q_{2B} - Q_{2A})$ with single temperature sink T_2. Thus it violates Kelvin-Planks law, so $\eta_B \geq \eta_A$ is true. $\eta_B = \eta_A$ means that HE_A is also a reversible heat engine. This leads to the important corollary of Carnot's theorem.

Corollary of Carnot's Theorem

All reversible engines working between the same fixed two temperatures have the same efficiency.

Proof: We have $\eta_B \geq \eta_A$ for B reversed

Thus $\eta_A \geq \eta_B$ for A reversed

so $\eta_A = \eta_B$ proved for both the engines are reversible.

7.5 ABSOLUTE THERMODYNAMIC TEMPERATURE SCALE OR KELVIN SCALE

We have $$\eta_{HE} = \frac{W_{net}}{Q_1} = \frac{Q_1 - Q_2}{Q_1} = 1 - \frac{Q_2}{Q_1}. \quad ...(7.7)$$

as $t_1 > t_2$

so $t_1 - t_2 > 0$ (must be for heat flow)

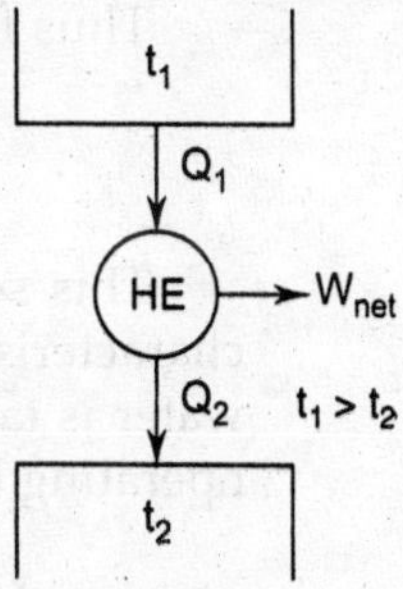

Fig. 7.4

We know reversible heat engine's efficiency depends on source and sink temperatures.

Thus $\eta_{rev} = f_1(t_1, t_2)$ where f_1 is a function of t_1 and t_2

From (7.7)

$$1 - \frac{Q_2}{Q_1} = f_1(t_1, t_2)$$

or $$\frac{Q_2}{Q_1} = 1 - f_1(t_1, t_2) \quad ...(7.8)$$

or $$\frac{Q_1}{Q_2} = \frac{1}{1 - f_1(t_1, t_2)} = f_2(t_1, t_2)$$

Let us now take three heat engines arranged as shown in Fig. 7.5

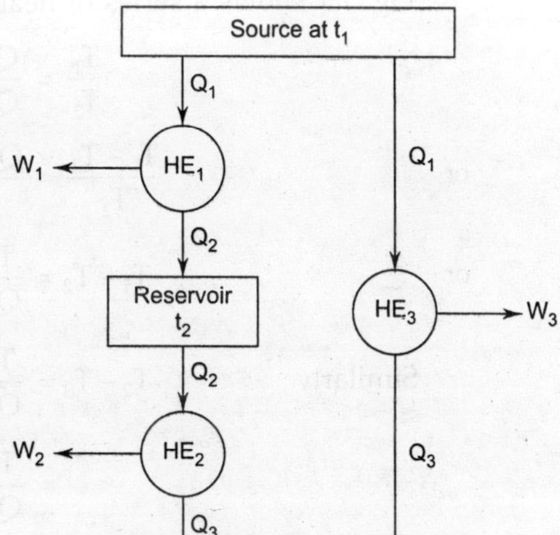

Fig. 7.5

$\therefore$ $$\frac{Q_2}{Q_3} = f_3(t_2, t_3) \quad ...(7.9)$$

and $$\frac{Q_1}{Q_3} = f_4(t_1, t_3) \quad ...(7.10)$$

Again

$$\frac{Q_1}{Q_2} = \frac{\dfrac{Q_1}{Q_3}}{\dfrac{Q_2}{Q_3}}$$

$$= \frac{f_4(t_1, t_3)}{f_3(t_2, t_3)}$$

or $$f_2(t_1, t_2) = \frac{f_4(t_1, t_3)}{f_3(t_2, t_3)} \quad ...(7.11)$$

Now t_1, t_2, t_3 are independent and t_3 does not affect the operation of HE_1 which is operating between t_1 and t_2.

To satisfy equation (7.11), we must have

$$f_4(t_1, t_3) = \phi(t_1)/\phi(t_3)$$

and $$f_3(t_2, t_3) = \phi(t_2)/\phi(t_3)$$

so, $$f_2(t_1, t_2) = \frac{\phi(t_1)}{\phi(t_2)} = \frac{Q_1}{Q_2} \quad ...(7.12)$$

Since $\phi(t)$ is an arbitrary function, the simpliest possible way to define absolute thermodynamic temperature T is to let $\phi(t)$ = T, proposed by Kelvin.

Thus from equation (7.12), we have

$$\frac{Q_1}{Q_2} = \frac{T_1}{T_2} \qquad ...(7.13)$$

This scale is also known as Kelvin scale. It is independent of characteristic of any substance. In Kelvin scale, the triple point of water is taken as the standard reference point. For a Carnot engine operating between T and T_{tr} (triple point of water), we have

$$\frac{Q}{Q_{tr}} = \frac{T}{T_{tr}} \qquad ...(7.14)$$

T_{tr} = 273.16 K (arbitrary assigned value)

or $$T = 273.16 \frac{Q}{Q_{tr}} \qquad ...(7.15)$$

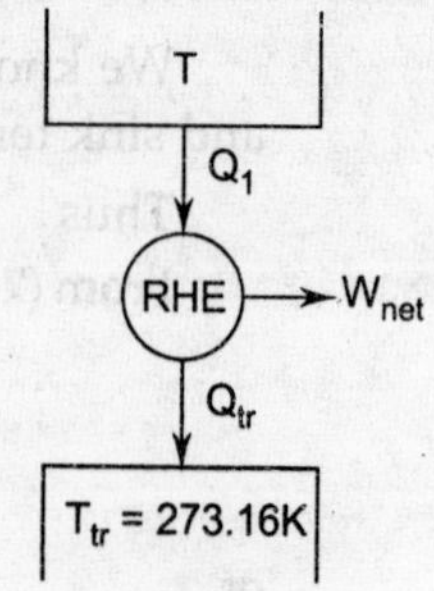

Fig. 7.6

If T changes, Q also changes. Thus Q plays an important role of thermodynamic property.

Fig. 7.7 shows a series of heat engines.

$$\frac{T_1}{T_2} = \frac{Q_1}{Q_2}$$

or $$\frac{T_1 - T_2}{T_2} = \frac{Q_1 - Q_2}{Q_2}$$

or $$T_1 - T_2 = \frac{T_2}{Q_2}(Q_1 - Q_2)$$

Similarly $$T_2 - T_3 = \frac{T_3}{Q_3}(Q_2 - Q_3)$$

$$= \frac{T_2}{Q_2}(Q_2 - Q_3)$$

$$T_3 - T_4 = \frac{T_2}{Q_2}(Q_3 - Q_4)$$

and so on.

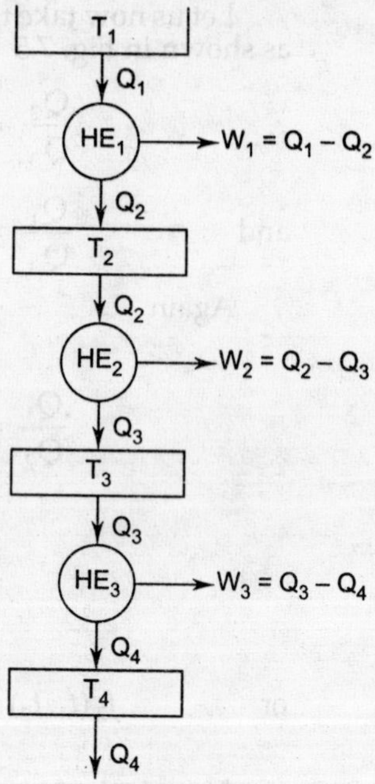

Fig. 7.7

Let $$W_1 = W_2 = W_3 = ... = W$$

$\therefore$ $$Q_1 - Q_2 = Q_2 - Q_3 = Q_3 - Q_4 = ...$$

Thus $$T_1 - T_2 = (T_2 - T_3) = (T_3 - T_4) = ... \qquad ...(7.16)$$

Also if it is assumed $Q_1 - Q_2 = Q_2 - Q_3 = Q_3 - Q_4 = ...$

then also $$W_1 = W_2 = W_3 = ... = W$$

Equation (7.16) shows equal temperature intervals usually steam point and ice point are taken as fixed reference temperatures and the difference between the two is arbitrary taken as 100°C.

$\therefore \quad T_s - T_i = 100$, [$T_s$ = steam point T_i = ice point] ...(7.17)

$$\eta_{rev} = 1 - \frac{T_i}{T_s} = 0.268 \text{ [}\eta_{rev} \text{ measured and found to be 0.268]}$$

or
$$\frac{T_i}{T_s} = 0.732 \quad ...(7.18)$$

(7.17) $\Rightarrow$ $T_s - 0.732\, T_s = 100$

or
$$T_s = \frac{100}{0.268} = 373.15 \text{ (steam point)}$$
$$T_i = 273.15 \text{ (ice point)}.$$

Triple point of water is at 0.01°C. For this in Kelvin scale 273.15 + 0.01 = 273.16 K.

7.6 ABSOLUTE ZERO ON THERMODYNAMIC TEMPERATURE SCALE

If more and more engines are added in series to Fig. 7.7, the total work output (W + W + W + ...) cannot exceed Q_1 (according to 1st law).

In the limiting case when total output is equal to Q_1, then the heat rejection to sink by the last engine will be zero. This will violate the Kelvin-Plank's 2nd law (some heat is to be rejected to sink). Therefore as the number of engines increases the heat rejection of the last engine decreases but it can never be zero. We can imagine that in the limiting case (no. of engines tends to infinity) the heat rejection is zero and the sink temperature in this case is also zero. Thus absolute zero is a conceptual lower limit on thermodynamic scale of temperature.

SOLVED EXAMPLES

Example 7.1. *Fig. 7.8 shows a heat pump (HP), a heat engine (HE). The heat pump is driven by heat engine. The work out from HE is also driving one machine giving 30KW work input. Find.*

(a) the rate of heat supply from the 840°C source i.e., Q_{1E} = ?

(b) the rate of total heat rejection to 60°C heat reservoir (sink).

i.e., $Q_{1P} + Q_{2E} = ?$

Solution.

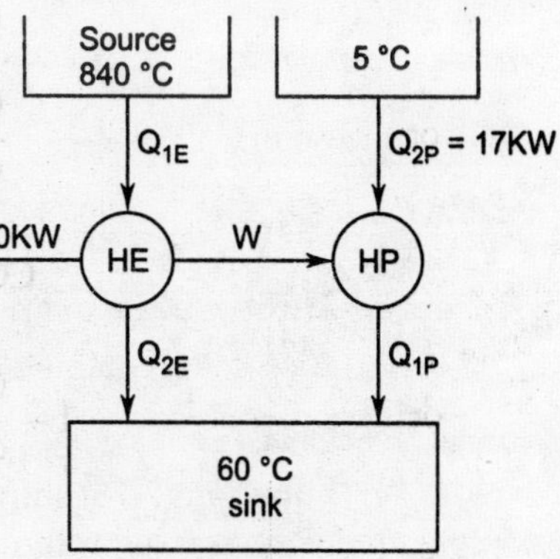

Fig. 7.8

$$\eta_{HP} = \frac{Q_{1P} - Q_{2P}}{Q_{1P}} = \frac{Q_{1P} - 17}{Q_{1P}}$$

or
$$1 - \frac{273 + 5}{273 + 60} = \frac{Q_{1P} - 17}{Q_{1P}}$$

$\therefore \quad Q_{1P} = 20.36$ KW

$[W = Q_{1P} \times \eta_{HP} = 20.36 \times 0.165 = 3.36 \text{ KW}]$

W_{net} for HE = 30 + W = 30 + 3.36

= 33.36 KW

for HE, $1-\dfrac{Q_{2E}}{Q_{1E}} = 1-\dfrac{60+273}{840+273}$

or $$\frac{Q_{2E}}{Q_{1E}} = 0.3 \quad \text{...(1)}$$

Again $$W_{net} = Q_{1E} - Q_{2E}$$

or $$33.36 = Q_{1E} - Q_{2E} \quad \text{...(2)}$$

Solving equations (1) and (2) we get $Q_{1E} = 47.66\text{KW}$. **Ans. (*a*)**

∴ $$Q_{2E} = 0.3 \times 47.66 = 14.3 \text{ KW}$$

Thus total heat rejection to sink at 60°C

$$= Q_{1P} + Q_{2E} = (20.36 + 14.3) \text{ KW}$$

$= 34.66$ KW. **Ans. (*b*)**

Example 7.2. *Fig. 7.9 shows a reversed Carnot heat engine (RCHE) (a refrigeration plant). Calculate the power required to drive the refrigeration plant.*

Solution.

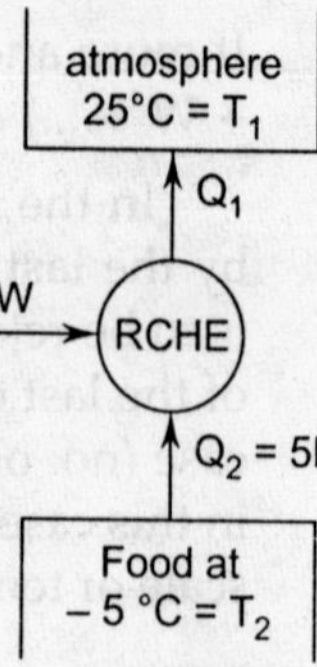

Fig. 7.9

We have $$\frac{Q_1}{Q_2} = \frac{T_1}{T_2}$$

or $$\frac{Q_1}{5} = \frac{273+25}{273-5}$$

or $$Q_1 = 5.56 \text{ KW}$$

∴ $$W = Q_1 - Q_2 = (5.56 - 5) = 0.56 \text{ KW}$$

= power required. **Ans.**

Example 7.3. *A heat engine is used to drive a heat pump. The heat transfers from the heat engine and from the heat pump are used to heat the water circulating through the radiators of a building. The thermal efficiency of the heat engine is 30% and the COP of the heat pump is 5. Evaluate the ratio of the total heat transfers to the circulating water to the heat input to heat engine.*

Solution. $$\eta_{HE} = 0.3$$

or $$1-\frac{Q_{2E}}{Q_{1E}} = 0.3$$

or $$\frac{Q_{2E}}{Q_{1E}} = 0.7 \quad \text{...(1)}$$

$$COP_{HP} = \frac{Q_{1P}}{Q_{1P} - Q_{2P}} = 5$$

or $$1-\frac{Q_{2P}}{Q_{1P}} = \frac{1}{5}$$

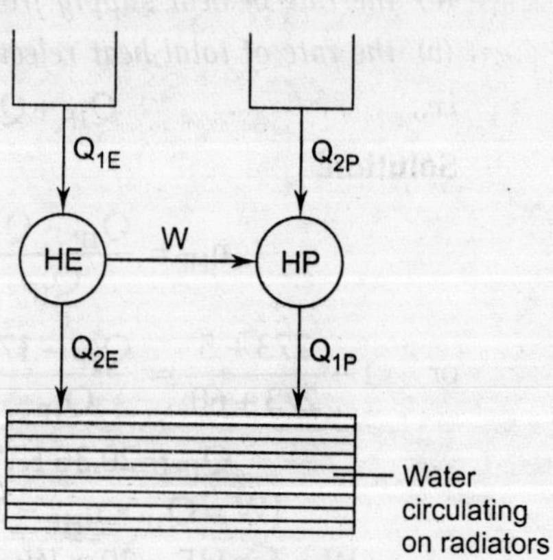

Fig. 7.10

or $$1-\frac{1}{5} = \frac{Q_{2P}}{Q_{1P}}, \text{ or } \frac{Q_{2P}}{Q_{1P}} = \frac{4}{5} \quad \text{...(2)}$$

We also have $$W = Q_{1E} - Q_{2E} = Q_{1P} - Q_{2P}$$

or $$Q_{1E} - Q_{1E} \times 0.7 = Q_{1P} - \frac{4}{5} Q_{1P}$$

or $$0.3\, Q_{1E} = \frac{Q_{1P}}{5}$$

or $$\frac{Q_{1P}}{Q_{1E}} = 1.5 \qquad ...(3)$$

To find $$\frac{Q_{2E} + Q_{1P}}{Q_{1E}} = \frac{Q_{2E}}{Q_{1E}} + \frac{Q_{1P}}{Q_{1E}} = 0.7 + 1.5$$

$$= 2.2$$

Thus the ratio is 2.2. **Ans.**

Example 7.4. *If 30 KJ are added to Carnot cycle at a temperature of 100°C and 20 KJ are rejected at 0°C, determine the location of absolute zero on the Celsius scale.*

Solution. $$\frac{Q_2}{Q_1} = \frac{T_2}{T_1} = \frac{x + 0}{x + 100}$$

or $$\frac{20}{30} = \frac{x}{x + 100}$$

or $$3x = 2x + 200$$

or $$x = 200$$

The location of absolute zero on the Celsius scale will be – 200°C.

Chapter **8**

Clausius Inequality, Entropy and Irreversibility

8.1 INTRODUCTION

In the first law, we have described a new property called internal energy. In second law we will come across another new property called entropy. As we have formulated mathematically the first law, the second law can also now be formulated mathematically after the introduction of the property entropy.

8.2 REVERSIBLE ADIABATIC PATHS DO NOT INTERSECT

Let us assume any two adiabatic paths CA & CB which intersect at C. Let us also draw one isothermal path which intersects CA and CB at A and B respectively. CA, AB and BC now constitute a cycle which is reversible. The work done on the system is the area included between them. During paths CA and CB, there is no heat transfer, the heat transfer takes place only the reversible isothermal path at a single temperature reservoir. So this cycle violates the 2nd law. Thus this reversible cycle is impossible.

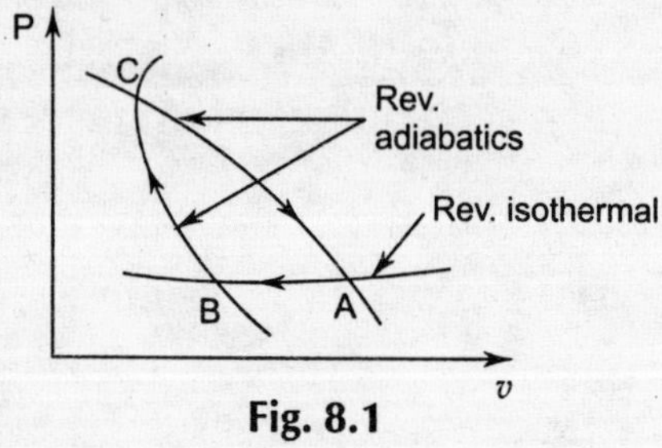

Fig. 8.1

So the starting assumption that two reversible adiabatic paths intersect is false. So two reversible adiabatic paths cannot intersect each other.

8.3 CLAUSIUS' THEOREM

Let a process be done reversibly from state 1 to state 2. Let reversible adiabatics $1 - a$ and $b - 2$ be drawn as show in Fig. 8.2. Then $a - b$ is drawn isothermally in such a manner that the area under 1 – *a* – *b* – 2 is same as the area under 1 – 2. Applying 1st law for process 1 – 2, we have

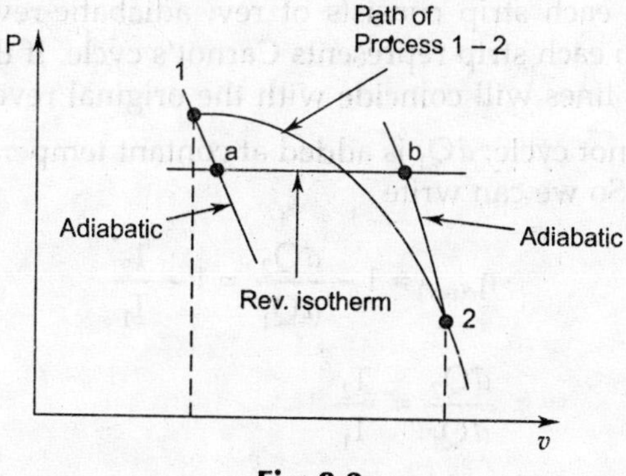

Fig. 8.2

$$Q_{1-2} = U_2 - U_1 + W_{1-2} \quad ...(8.1)$$

for process $(1 - a - b - 2)$

$$Q_{1-a-b-2} = U_2 - U_1 + W_{1-a-b-2} \quad ...(8.2)$$

Since $W_{1-2} = W_{1-a-b-2}$, we can write

$$Q_{1-2} = Q_{1-a-b-2}$$

$$= Q_{1-a} + Q_{a-b} + Q_{b-2} = 0 + Q_{a-b} + 0$$

or

$$Q_{1-2} = Q_{a-b} \quad ...(8.3)$$

Equation (8.3) states that heat transferred in the process 1 – 2 is equal to heat transferred in the isothermal process *a* – *b*.

So any reversible path can be replaced by a zig-zag path between the same end states. The zig-zag path consists of a reversible adiabatic followed by a reversible isotherm and then by a reversible adiabatic in such a way that will satisfy equation (8.3) *i.e.*, the heat transferred during the isothermal process is the same as that transferred during the original process.

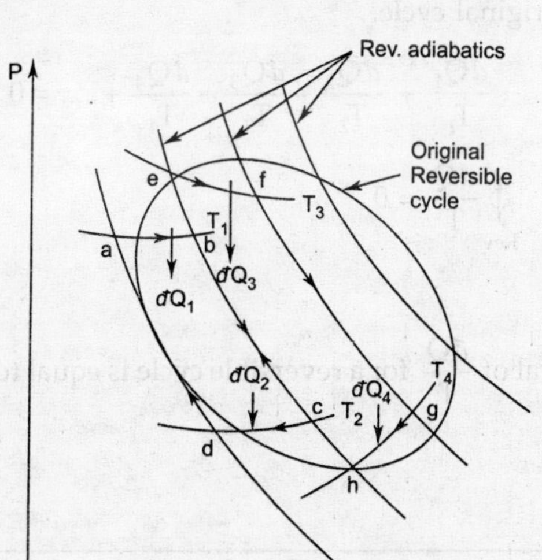

Fig. 8.3

Fig. 8.3 shows an original reversible cycle which is divided by number of strips of reversible adiabatic lines. The strips are closed on tops and bottoms by means of reversible isotherms. Thus each strip consists of rev. adiabatic-rev. isotherm-rev. adiabatic-rev. isothern lines. So each strip represents Carnot's cycle. If the no. of strips are very large, then the zig-zag lines will coincide with the original reversible cycle.

For *abcd* Carnot cycle; $đQ_1$ is added at contant temperature T_1 and $đQ_2$ is rejected at temperature T_2. So we can write

$$\eta_{c(rev)} = 1 - \frac{đQ_2}{đQ_1} = 1 - \frac{T_2}{T_1}$$

or $$\frac{đQ_2}{đQ_1} = \frac{T_2}{T_1}$$

or $$\frac{đQ_1}{đQ_2} = \frac{T_1}{T_2}$$

$$\frac{đQ_1}{T_1} = \frac{đQ_2}{T_2}$$, He have $đQ_1$ is positive and $đQ_2$ is negative

Thus $$\frac{đQ_1}{T_1} = -\frac{đQ_2}{T_2}$$ (taking proper sign of heat)

or $$\frac{đQ_1}{T_1} + \frac{đQ_2}{T_2} = 0 \qquad ...(8.4)$$

Similarly for the elemental Carnot cycle *efgh* we have

$$\frac{đQ_3}{T_3} + \frac{đQ_4}{T_4} = 0 \qquad ...(8.5)$$

Thus applying the above procedure for all the elemental Carnot's cycles, we get the equation for whole original cycle.

Thus $$\frac{đQ_1}{T_1} + \frac{đQ_2}{T_2} + \frac{đQ_3}{T_3} + \frac{đQ_4}{T_4} + ... = 0$$

or, $$\oint_{Rev} \frac{đQ}{T} = 0 \qquad ...(8.6)$$

Thus

"The cyclic integral of $\frac{đQ}{T}$ for a reversible cycle is equal to zero". This is the Clausius theorem.

8.4 ENTROPY

Let process 1 – 2 is achieved through reversible path R_1 and the cycle is completed by process 2 – 1 through reversible path R_2.

Applying Clausius' theorem to paths R_1 & R_2, we get

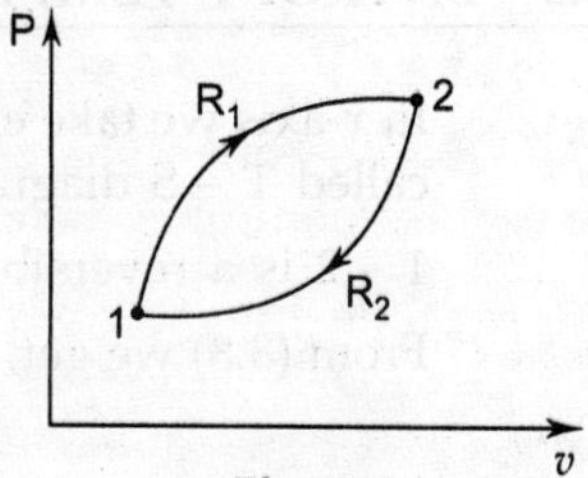

Fig. 8.4

$$\oint_{1\,R_1}^{2} \frac{đQ}{T} + \oint_{2\,R_2}^{1} \frac{đQ}{T} = 0$$

or $$\int_{1\,R_1}^{2} \frac{đQ}{T} - \int_{1\,R_2}^{2} \frac{đQ}{T} = 0$$

or $$\int_{1\,R_1}^{2} \frac{đQ}{T} = \int_{1\,R_2}^{2} \frac{đQ}{T} \qquad ...(8.7)$$

We have taken arbitrarily any two reversible paths and we have come upto an equation which shows that the expression $\int_{1\,R}^{2} \frac{đQ}{T}$ is independent of reversible path connecting states 1 and 2. Thus there exists a property $\frac{đQ_R}{T}$ which is called entropy (S).

Thus entropy is a point function.

so $$\int_{1\,R}^{2} \frac{đQ}{T} = S_2 - S_1$$

When two states 1 – 2 are very near, then

$$\frac{đQ_R}{T} = dS. \qquad ...(8.8)$$

Where dS is the change of entropy. ds is an exact differential because S is a point function and a thermodynamic property. $đQ_R$ heat transfer is reversible. (R → Reversible).

Entropy is taken from Greek Word 'tropee' meaning transformation and introduced by Clausius. Entropy per unit mass is called *specific entropy*. So mathematically

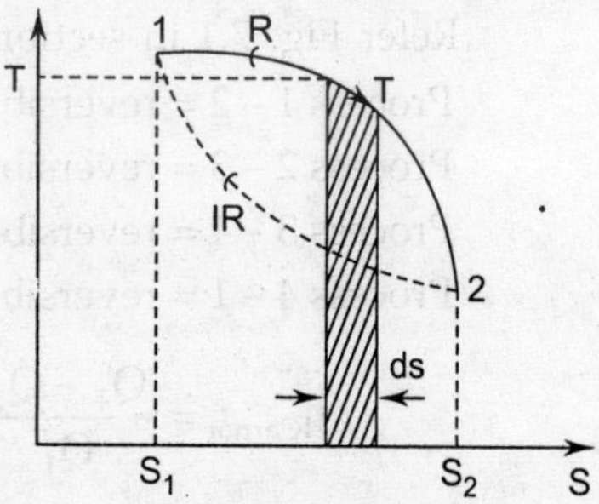

Fig. 8.5

specific entropy $$s = \frac{S}{m} \text{ J/kg K.}$$

In Fig. 8.5

R = Reversible path

IR = Irreversible path

IR is replaced by R in view point of entropy which depends on end points. We integrate in the reversible path to get entropy for irreversible path.

$$\therefore \quad S_2 - S_1 = \int_1^2 \frac{đQ_R}{T} = \sum_1^2 \delta S_{IR}.$$

Integration is performed only on a reversible path.

8.5 ENTROPY-TEMPERATURE PLOT

In x-axis we take entropy (s) and in y-axis we take Kelvin temperature (T). It is generally called T – S diagram.

1 – 2 is a reversible process

From (8.8) we get,

$$dS = \frac{đQ_R}{T}$$

or $$đQ_R = T\,dS$$

or $$Q_R = \int_1^2 T\,dS \qquad ...(8.9)$$

= area under the curve 1 – 2

= Heat transfer in the process.

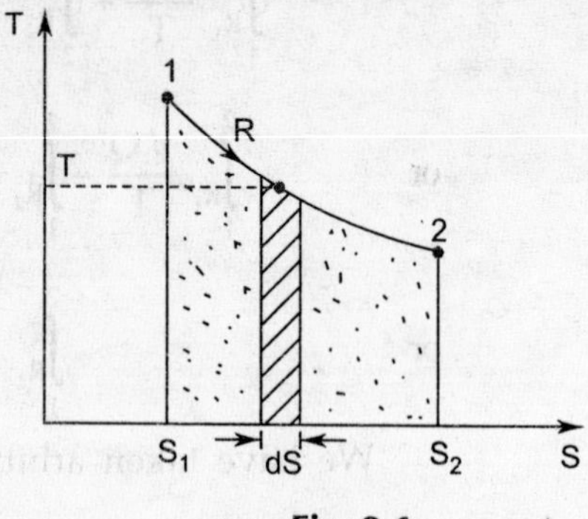

Fig. 8.6

8.5.1 Reversible Adiabatic Process

If 1 – 2 process is reversible adiabatic, then $đQ_R = 0$

or from (8.8) we have $dS = \frac{đQ_R}{T} = 0$

or $S = C = \text{constant } i.e., S_2 = S_1$

Thus a reversible adiabatic process is an isentropic process.

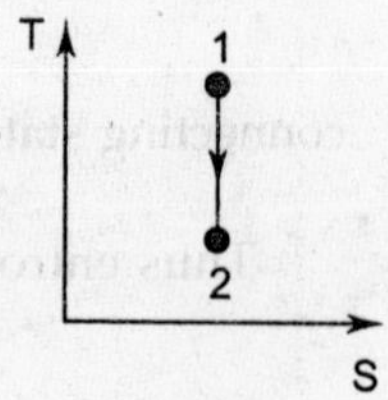

Fig. 8.7

8.5.2 Reversible Isothermal Process

$$Q_R = \int_1^2 T\,dS = T\int_1^2 dS$$

$$= T(S_2 - S_1) \qquad ...(8.10)$$

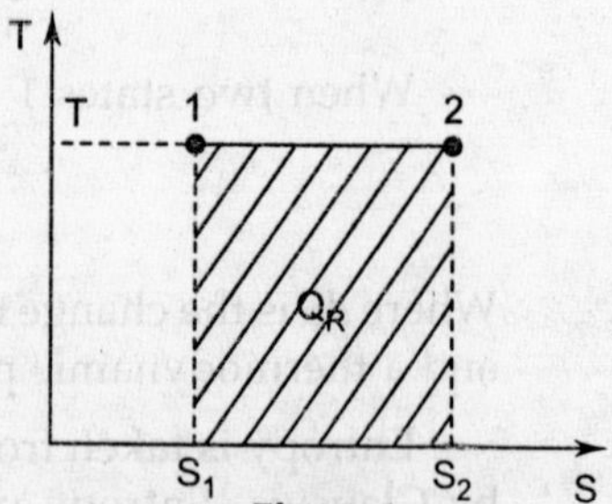

Fig. 8.8

8.5.3 Carnot Cycle in T–S Plot

Refer Fig. 7.1 in section 7.1 of chapter 7, we have the following process.

Process 1 – 2 = reversible isothermal

Process 2 – 3 = reversible adiabatic

Process 3 – 4 = reversible isothermal

Process 4 – 1 = reversible adiabatic

$$\eta_{Carnot} = \frac{Q_1 - Q_2}{Q_1} = \frac{T_1(S_2 - S_1) - T_2(S_3 - S_4)}{T_1(S_2 - S_1)}$$

$$= \frac{T_1(S_2 - S_1) - T_2(S_2 - S_1)}{T_1(S_2 - S_1)}$$

$$= \frac{(T_1 - T_2)(S_2 - S_1)}{T_1(S_2 - S_1)} = \frac{T_1 - T_2}{T_1} = 1 - \frac{T_2}{T_1}.$$

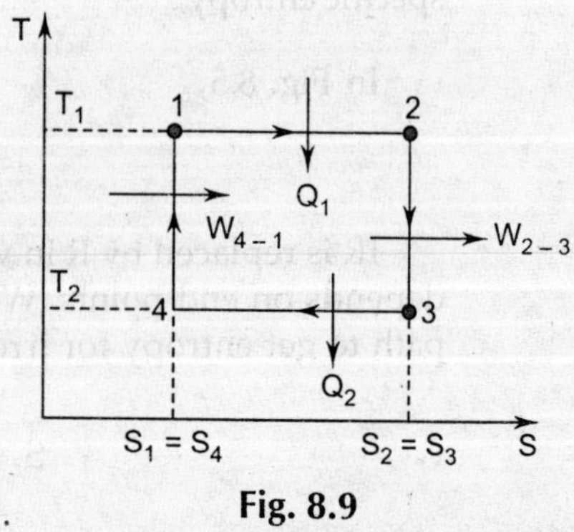

Fig. 8.9

8.6 CLAUSIUS'S INEQUALITY

Let ABCD is a cyclic process where, AB is a general process (reversible or irreversible) and other processes are reversible. Now ABCD is divided into a number of small cycles. Let us take AEFD as one of these small cycles and its efficiency is $\eta = 1 - \frac{đQ_2}{đQ}$ where $đQ$ is heat added to isothermal (T = C) process AE and $đQ_2$ is the rejection at T_2 in the isothermal process FD.

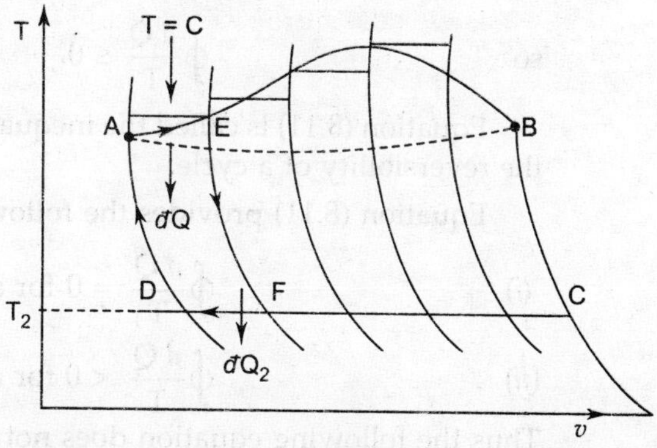

Fig. 8.10

We know that

$$\eta_{\text{general}} \le \eta_{\text{rev}}$$

or
$$1 - \frac{đQ_2}{đQ} \le \left(1 - \frac{đQ_2}{đQ}\right)_{\text{rev}}$$

or
$$\frac{đQ_2}{đQ} \ge \left(\frac{đQ_2}{đQ}\right)_{\text{rev}}$$

or
$$\frac{đQ}{đQ_2} \le \left(\frac{đQ}{đQ_2}\right)_{\text{rev}}$$

Since
$$\left(\frac{đQ}{đQ_2}\right)_{\text{rev}} = \frac{T}{T_2},$$

thus
$$\frac{đQ}{đQ_2} \le \frac{T}{T_2}$$

or
$$\frac{đQ}{T} \le \frac{đQ_2}{T_2} \quad \text{for any general process AB, (reversible or irreversible)}$$

For a reversible process $dS = \frac{đQ_{\text{rev}}}{T} = \frac{đQ_2}{T_2}$.

Hence for any process $\frac{đQ}{T} \le dS$ ...(8.10(*a*))

Thus in any cyclic process

$$\oint \frac{đQ}{T} \le \oint dS.$$

Since entropy is a point function and in a cyclic process initial and final point is the same, thus $$\oint dS = 0$$

so $$\oint \frac{đQ}{T} \leq 0. \qquad ...(8.11)$$

Equation (8.11) is called the inequality of Clausius which supplies information about the reversibility of a cycle.

Equation (8.11) provides the following two equations

(*i*) $$\oint \frac{đQ}{T} = 0 \text{ for a reversible cycle.}$$

(*ii*) $$\oint \frac{đQ}{T} < 0 \text{ for a irreversible cycle (Possible).}$$

Thus the following equation does not exist *i.e.,*

$$\oint \frac{đQ}{T} > 0 \text{ impossible and also violates second law.}$$

8.7 ENTROPY CHANGE IN AN IRREVERSIBLE PROCESS

From equation (8.10*a*) we can write for any process

$$\frac{đQ}{T} \leq dS$$

or $$dS \geq \frac{đQ}{T} \qquad ...(8.12)$$

For clarification of equation (8.12), let us consider Fig. 8.11, where *a* and *b* are reversible process and *c* is an irreversible process.

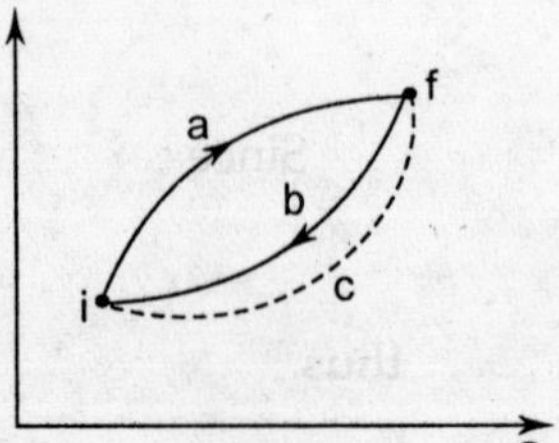

Fig. 8.11

For (*a*) & (*b*) cycle

$$\oint_{Rev} \frac{đQ}{T} = \int_{i(a)}^{f} \frac{đQ}{T} + \int_{f(b)}^{i} \frac{đQ}{T} = 0$$

or $$\int_{i(a)}^{f} \frac{đQ}{T} = -\int_{f(b)}^{i} \frac{đQ}{T} \qquad ...(8.13)$$

For (*a*) & (*c*) cycle

$$\oint \frac{đQ}{T} = \int_{i(a)}^{f} \frac{đQ}{T} + \int_{f(c)}^{i} \frac{đQ}{T} < 0 \text{ (for } a \,\&\, c \text{ irreversible cycle)} \qquad ...(8.14)$$

From equation (8.13) & (8.14) we can write

$$-\int_{i(a)}^{f} \frac{đQ}{T} + \int_{f(c)}^{i} \frac{đQ}{T} < 0$$

or $$\int_{f(b)}^{i} \frac{đQ}{T} > \int_{f(c)}^{i} \frac{đQ}{T} \qquad ...(8.15)$$

or $$\int_{f(b)}^{i} dS > \int_{f(c)}^{i} \frac{đQ}{T} \quad \text{(as } b \text{ is reversible)}$$

or $$\int_{f(c)}^{i} dS > \int_{f(c)}^{i} \frac{đQ}{T} \quad \text{(as entropy change is same in } b \text{ and } c\text{)}$$

So for any irreversible process $dS > \frac{đQ}{T}$ and for a reversible process $dS = \frac{đQ_{rev}}{T}$.

Therefore, for a general process, we can write

$$dS \geq \frac{đQ}{T}$$

or $$S_f - S_i \geq \int_i^f \frac{đQ}{T}. \qquad ...(8.16)$$

Note that = sign holds good for reversible process and > sign for irreversible process.

8.8 PRINCIPLE OF INCREASE OF ENTROPY

For any process we can write for the total mass

$$dS \geq \frac{đQ}{T}.$$

If the system is isolated, $đQ = 0$

$\therefore$ $(dS)_{isol} \geq 0$

For reversible process $(dS)_{isol} = 0$ ...(8.17)

or $S = \text{const.}$

For irreversible process $(dS)_{isol} > 0$ ....(8.18)

Thus it is proved that the entropy of an isolated system always increases, it can never decrease. The last two lines mentioned above is called the principle of increase of entropy or simply entropy principle and this is the general quantitative statement of 2nd law.

We can form an isolated system by putting system and surroundings together. The universe is an isolated system.

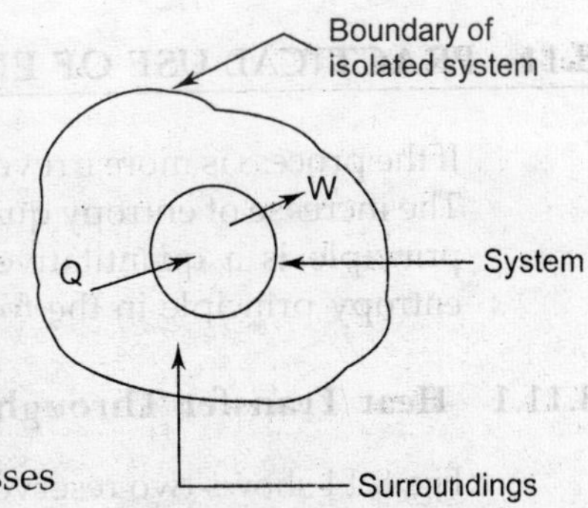

Fig. 8.12

Universe = System + surroundings

$\therefore$ $dS_{univ} \geq 0$ for all possible processes

or $dS_{system} + dS_{surr} \geq 0.$

The entropy may decrease locally at some region within the isolated system, but it will be compensated by a greater increase of entropy somewhere within the system. So that the net effect of an irreversible process will be an increase of entropy of the whole system.

8.9 THE DEGREE OF IRREVERSIBILITY OF AN IRREVERSIBLE PROCESS

If a cyclic process consisting of different irreversible processes, then as per Clausius inequality

$$\oint \frac{đQ}{T} < 0$$

or
$$\oint \frac{đQ}{T} + I = 0 \qquad ...(8.19)$$

where I is known as measure of irreversibility, it satisfies $I \geq 0$ for a cycle.

8.10 SUMMARY OF FIRST AND SECOND LAW BY CLAUSIUS

Clausius summarized the thermodynamics laws 1st & 2nd in the following sentences.

(*a*) The energy of the universe remains constant.

(*b*) The entropy of the universe tends towards a maximum.

In an isolated system, the entropy always increases and it is maximum at equilibrium state.

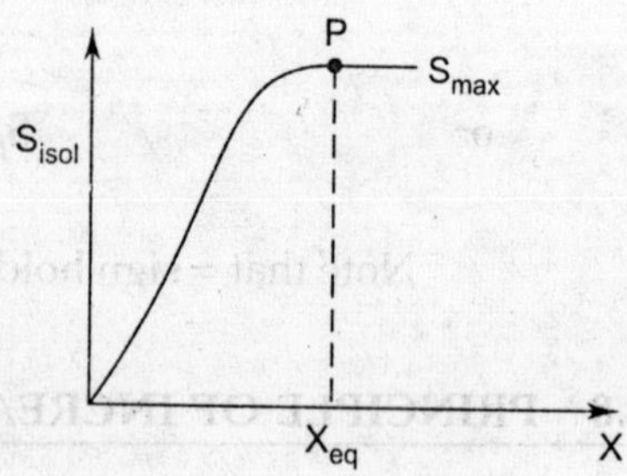

Fig. 8.13

$$S_{isol} = f(x)$$

where x is a parameter.

$$S_{isol} = S_{max}$$

at a point $x = x_{eq}$

where x_{eq} is the value of the parameter x at equilibrium state

So $\frac{dS}{dx} = 0$ at $x = x_{eq}$ represents the equilibrium state.

Thus $dS = 0$ at equilibrium state, it means at equilibrium state any conceivable change in entropy would be zero.

8.11 PRACTICAL USE OF ENTROPY PRINCIPLE

If the process is more irreversible, the more will be the increase of entropy of the universe. The increase of entropy quantifies the extent of irreversibility of the process. The entropy principle is a quantitative statement of 2nd law. We illustrate some applications of entropy principle in the following.

8.11.1 Heat Transfer Through a Finite Temperature Difference

Fig. 8.14 shows two reservoirs A and B with T_1 & T_2 temperatures respectively. Let $T_1 > T_2$, so heat (Q) will flow from A to B. So for the isolated system comprising of two reservoirs A & B and the rod, we can write

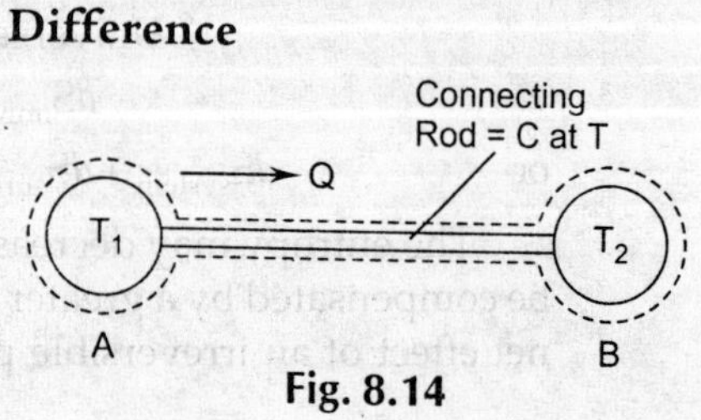

Fig. 8.14

$$\Delta S_{univ} = \Delta S_A + \Delta S_B + \Delta S_C$$

$$= -\frac{Q}{T_1} + \frac{Q}{T_2} + \frac{-Q+Q}{T} \qquad [\textit{i.e.}, \Delta S_C \text{ is zero at steady state}]$$

$$= \frac{Q(T_1 - T_2)}{T_1 T_2} + 0$$

$$= Q\frac{(T_1 - T_2)}{T_1 T_2}$$

Since $T_1 > T_2$, $\Delta S_{univ} > 0$ and the process is possible and irreversible. If $T_1 = T_2$, $\Delta S_{univ} = 0$, the process is reversible.

If $T_2 > T_1$, $\Delta S_{univ} < 0$ and the process is impossible.

8.11.2 Mixing of two Fluids

Subsystem A	*Subsystem B*
mass m_1 specific heat c_1 temperature T_1	m_2 c_2 T_2.

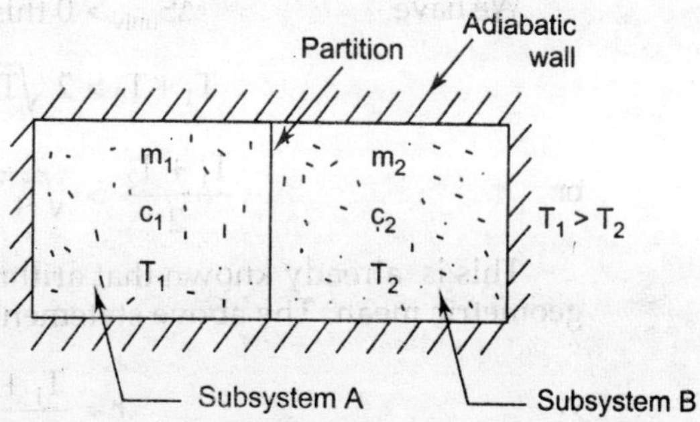

Fig. 8.15

There are two subsystems A and B as shown in Fig. 8.15. When the partition is removed, the two fluids mix together and at equilibrium let T_f be the final temperature. So, $T_2 < T_f < T_1$. As there is no heat transfer through adiabatic wall, we have

$$\text{Heat lost by A} = \text{Heat gained by B}$$

$$m_1c_1(T_1 - T_f) = m_2c_2(T_f - T_2)$$

or,
$$T_f = \frac{m_1 c_1 T_1 + m_2 c_2 T_2}{m_1 c_1 + m_2 c_2} \qquad ...(8.20)$$

The change of entropy for subsystem A is

$$\Delta S_1 = \int_{T_1}^{T_f} \frac{đQ_{rev}}{T} = \int_{T_1}^{T_f} \frac{m_1 c_1 dT}{T} = m_1c_1 \ln \frac{T_f}{T_1}.$$

$\Delta S_1 < 0$, since $T_f < T_1$.

The change of entropy for subsystem B is

$$\Delta S_2 = \int_{T_1}^{T_f} \frac{đQ_{rev}}{T} = \int_{T_2}^{T_f} \frac{m_2 c_2 dT}{T} = m_2c_2 \ln \frac{T_f}{T_2}.$$

As $T_f > T_2$, $\Delta S_2 > 0$.

$$\therefore \quad \Delta S_{univ} = \Delta S_1 + \Delta S_2$$

$$= m_1c_1 \ln \frac{T_f}{T_1} + m_2c_2 \ln \frac{T_f}{T_2} \qquad ...(8.21)$$

In actual calculation the RHS of equation (8.21) is positive,

Thus $\Delta S_{univ} > 0$ and the mixing process is irreversible. For integration the irreversible path is replaced by reversible paths.

For $m_1 = m_2 = m$ (say), and $c_1 = c_2 = c$ (say)

$$\Delta S_{univ} = mc \ln \frac{T_f^{\,2}}{T_1 T_2}$$

where $$T_f = \frac{m_1 c_1 T_1 + m_2 c_2 T_2}{m_1 c_1 + m_2 c_2} = \frac{T_1 + T_2}{2}$$

Thus $$\Delta S_{univ} = 2mc \ln \left(\frac{T_1 + T_2}{2\sqrt{T_1 T_2}} \right).$$

We have $\Delta S_{univ} > 0$ this is possible when

$$T_1 + T_2 > 2\sqrt{T_1 T_2}.$$

or $$\frac{T_1 + T_2}{2} > \sqrt{T_1 T_2}.$$

This is already known that arithmetic mean of any two numbers is greater than the geometric mean. The above statement can be proved geometrically as

$$\therefore \quad r = \frac{T_1 + T_2}{2}$$

$$r_1^2 = T_1 T_2$$

or $$r_1 = \sqrt{T_1 T_2}$$

$$r > r_1$$

or $$\frac{T_1 + T_2}{2} > \sqrt{T_1 T_2} \textbf{ proved.}$$

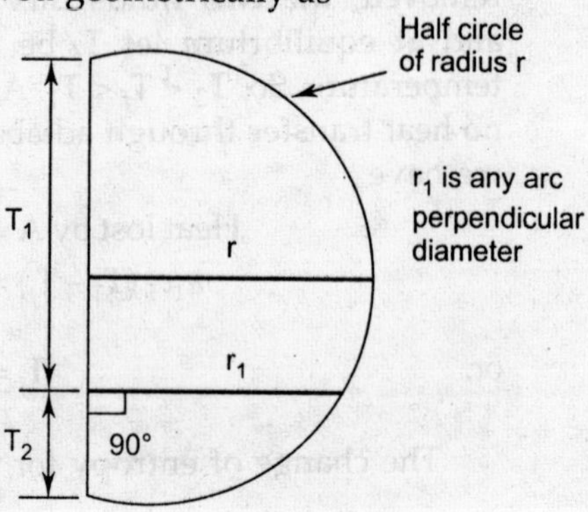

Fig. 8.15(a)

The reverse of the statement *i.e.*, if

$$\frac{T_1 + T_2}{2} > \sqrt{T_1 T_2} \text{ the } \Delta S_{univ} > 0.$$

8.11.3 Maximum Work Out (from two Fixed Temperatures) by an Engine Working Reversibly

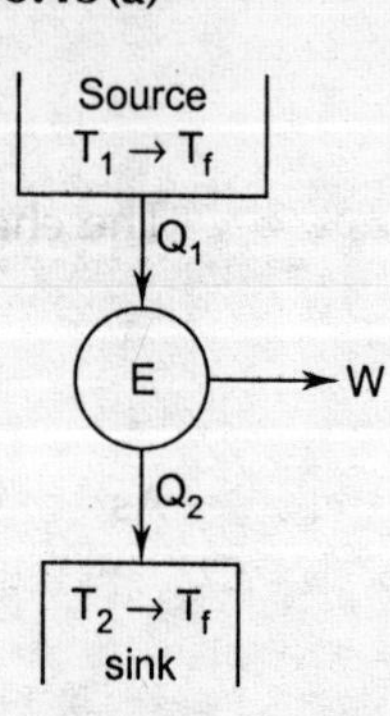

Fig. 8.16

$$\Delta S_{univ} = 0$$

or $$\Delta S_1 + \Delta S_2 = 0$$

or $$mc_p \int_{T_1}^{T_f} \frac{dT_1}{T_1} + mc_p \int_{T_2}^{T_f} \frac{dT_2}{T_2} = 0$$

or $$\int_{T_1}^{T_f} \frac{dT_1}{T_1} + \int_{T_2}^{T_f} \frac{dT_2}{T_2} = 0$$

or $$\ln \frac{T_f}{T_1} + \ln \frac{T_f}{T_2} = 0$$

or $$\ln \frac{T_f^2}{T_1 T_2} = 0 = \ln 1$$

or $$T_f = \sqrt{T_1 T_2}\,.$$

When the system reaches T_f, the workdone by the engine during that period is given by

$$\begin{aligned} W_{max} &= \text{Heat lost by source} - \text{Heat gained by sink} \\ &= Q_1 - Q_2 \\ &= mc_p(T_1 - T_f) - mc_p(T_f - T_2) \\ &= mc_p[(T_1 + T_2) - 2T_f] \\ &= mc_p\left[(T_1 + T_2) - 2\sqrt{T_1 T_2}\right] \\ &= mc_p\left(\sqrt{T_1} - \sqrt{T_2}\right)^2 \end{aligned}$$

c_p is the specific heat at constant pressure in (KJ/*kg* K).

8.11.4 External Mechanical Irreversibility

(*i*) Isothermal dissipation of work

Let I current flows through a resistor in contact with reservoir. At steady state,

$$dU = 0,$$

∴ $$Q = dU + W$$

or $$-Q = 0 + (-W)$$

$dU = 0$ for isothermal process

or $$\boxed{Q = W}$$

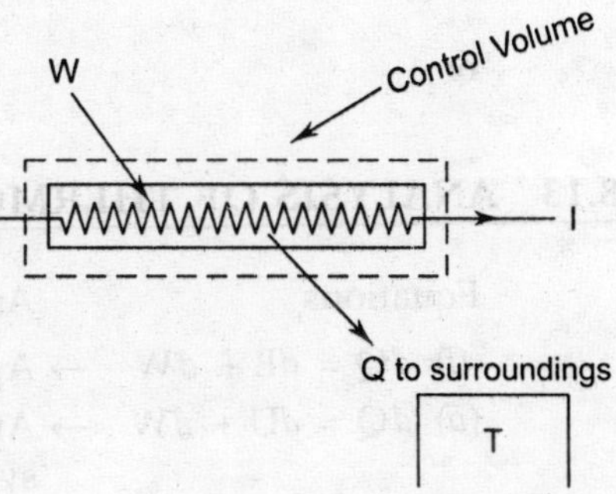

Fig. 8.17

$$\Delta S_{surr} = \frac{Q}{T} = \frac{W}{T}$$

At steady state $\Delta S_{system} = 0$

or $$\begin{aligned} \Delta S_{univ} &= \Delta S_{sys} + \Delta S_{surr} \\ &= 0 + \Delta S_{surr} \\ &= \Delta S_{surr} = \frac{W}{T} \end{aligned}$$

∴ $$\Delta S_{univ} > 0 \text{ as } W > 0.$$

Thus there is increase of entropy of the universe from this irreversible process.

(*ii*) Adiabatic dissipation of work

Let W is the work transfer to the system, according to 1st law,

$$đQ = dU + đW$$

or $$0 = dU + đW$$

or $$U_2 - U_1 = -\int đW = -W$$

$$\Delta S_{univ} = \Delta S_{surr} + \Delta S_{sys}.$$

$$= 0 + mc_p \ln \frac{T_f}{T_i}$$

where c_p is the specific heat at constant pressure.

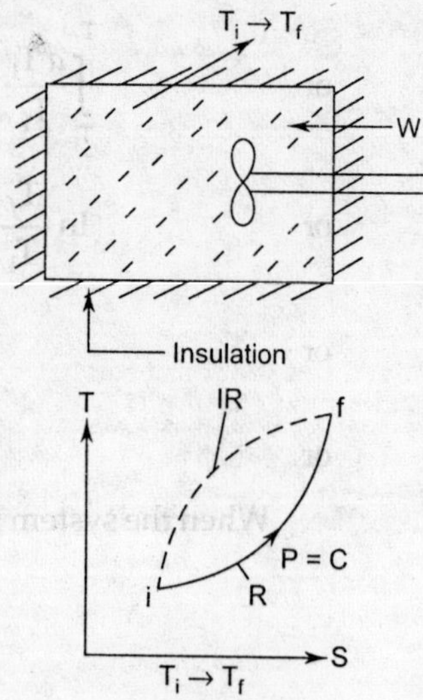

Fig. 8.18

8.12 FIRST LAW AND SECOND LAW COMBINED

From 1st law we get $đQ = dU + PdV$

From 2nd law we get $đQ_{rev} = TdS$

∴ Combining 1st & 2nd law

$$TdS = dU + PdV \quad \text{...(8.22)}$$

Again for enthalpy $H = U + PV$

or $$dH = dU + PdV + VdP$$

$$= TdS + VdP$$

or $$TdS = dH - VdP \quad \text{...(8.23)}$$

8.13 ANALYSIS OF THERMODYNAMIC EQUATIONS

Equations	Analysis
(*i*) $đQ = dE + đW$	→ Applicable for any process.
(*ii*) $đQ = dU + đW$	→ Applicable for any process undergone by a closed stationary system.
(*iii*) $đQ = dU + PdV$	→ Applicable for closed system when PdV is the only work present. True for reversible process.
(*iv*) $đQ_{rev} = TdS$	→ True for reversible process.
(*v*) $TdS = dU + PdV$	→ Applicable for any process in closed system.
(*vi*) $TdS = dH - VdP$	→ Applicable for any closed system.

SOLVED EXAMPLES

Example 8.1. *In a process friction causes the temperature rise of water from 40°C to 43°C. The process is adiabatic what is the change of entropy per kg of water flow at constant volume? [C (water) = 4.187 KJ/kg k]*

Solution: $đQ_{rev} = TdS$

or $$dS = \frac{đQ}{T} = \frac{mc\,dT}{T}$$

or $$\int dS = \int_{(40+273)}^{(43+273)} \frac{mc\,dT}{T}$$

$$= 1 \times 4.187 \times \ln\left(\frac{43+273}{40+273}\right)$$

$$S_2 - S_1 = 0.04\ \text{KJ/K} \quad \textbf{Ans.}$$

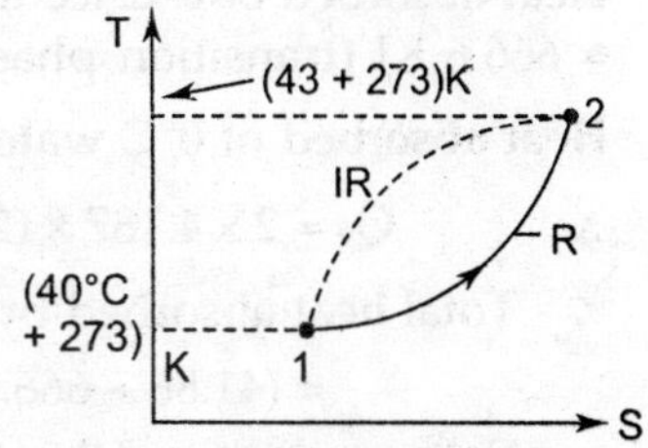

Fig. 8.19

Example 8.2. *Two kgs of water at 300 K is brought into contact with heat reservoir at 400 K. When the water has reached 400 K, find the entropy change for (a) water (b) heat reservoir (c) universe.*

Solution.

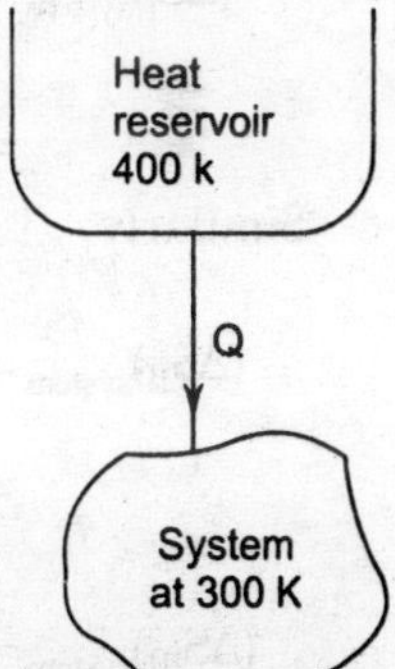

Fig. 8.20

(*a*) $$(\Delta S)_{water} = \int \frac{đQ}{T} = \int_{T_1=300}^{T_2=400} \frac{mc_v\,dT}{T}$$

$$= mc_v \ln\frac{T_2}{T_1} = 2 \times 4.187 \ln\frac{400}{300}$$

$$= 2.41\ \text{KJ/K} \quad \textbf{Ans.}$$

(*b*) Heat $$Q = mc_v(T_2 - T_1)$$
$$= 2 \times 4.187\,(400 - 300)$$
$$= 837.4\ \text{KJ}$$

For reservoir

$$\therefore \quad (\Delta S)_{res} = \frac{-Q}{T} = \frac{-837.4}{400} = -2.09\ \text{KJ/K} \quad \textbf{Ans.}$$

(*c*) Change of entropy of the universe

$$(\Delta S)_{univ} = (\Delta S)_{\substack{sys \\ (=water)}} + (\Delta S)_{\substack{surr \\ (=res)}}$$

$$= 2.41 - 2.09$$
$$= 0.32\ \text{KJ/K} \quad \textbf{Ans.}$$

Example 8.3. *Two kgs of ice at -10°C is exposed to the atmosphere which is at 25°C. The ice melts and comes into thermal equilibrium with the atmosphere.*
Determine the entropy increase of the universe.

$$[c_{ice} = 2.093,\ L_{ice} = 333.3,\ C_{water} = 4.187]$$

Solution.

Heat absorbed of *ice* at – 10°C to become 0°C *ice* = $Q_1 = mc_{ice}\,(0 + 10)$

$$= 2 \times 2.093 \times 10 = 41.86\ \text{KJ (at solid phase)}$$

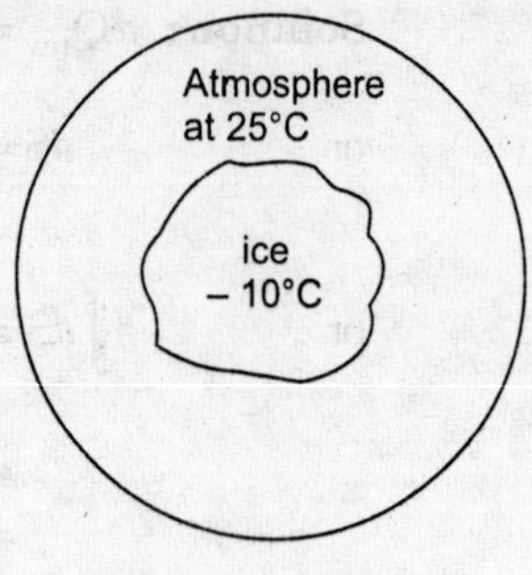

Fig. 8.21

Heat absorbed of 0°C ice to become 0°C water $Q_2 = 2 \times 333.3$
= 666.6 KJ (transition phase)

Heat absorbed of 0°C water to become 25°C water

$\therefore \quad Q_3 = 2 \times 4.187 \times (25 - 0) = 209.35$ KJ

$\therefore$ Total heat absorbed by ice 'Q' from atmosphere

$= (41.86 + 666.6 + 209.35)$ KJ = 917.81 KJ

$\therefore$ Entropy change of the atmosphere

$$= (\Delta S)_{atm} = \frac{-Q}{T} = \frac{-917.81}{273+25} = -3.08 \text{ KJ/K}$$

Entropy change of the system (*ice*) as it gets heated from – 10°C to 0°C

$$(\Delta S_I)_{system} = \int_{263}^{298} mc_p \frac{dT}{T} = 2 \times 2.093 \ln \frac{298}{263}$$

$$= 0.523 \text{ KJ/K}$$

Similarly

$$(\Delta S_{II})_{system} = \frac{333.3 \times 2}{273} = 1.22 \times 2 \text{ KJ/K}$$

$$= 2.44 \text{ KJ/K}$$

$$(\Delta S_{III})_{system} = \int_{273}^{298} mc_p \frac{dT}{T}$$

$$= 2 \times 4.187 \ln \frac{298}{273}$$

$$= 0.733 \text{ KJ/K}$$

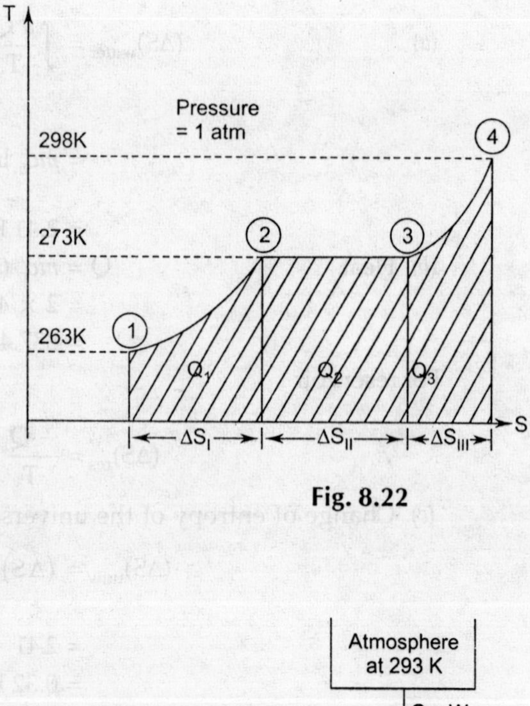

Fig. 8.22

Total entropy change of ice as it melts to water at 25°C

$$= (0.523 + 2.44 + 0.733)$$

$$= 3.696 \text{ KJ/K}$$

$$\therefore (\Delta S)_{univ} = (\Delta S)_{system} + (\Delta S)_{atm}$$

$$= 3.696 + (-3.08)$$

$$= 0.616 \text{ KJ/K} \quad \textbf{Ans.}$$

Example 8.4. *What is the minimum amount of work necessary to convert one kg of water at 20°C into ice at – 5°C? Take c_p of ice = 2.093 KJ/kgk,*

Latent heat of fusion of ice = 333.3 KJ/kg,

c_{water} = 4.187 KJ/kg K.

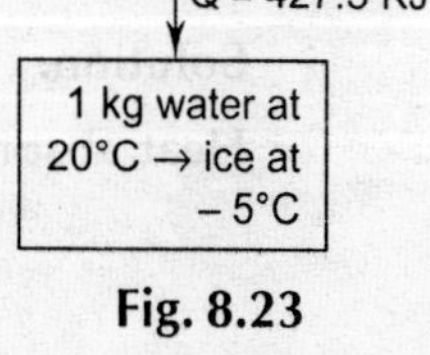

Fig. 8.23

Solution. Heat removal from water at 20°C to convert it to *ice* at – 5°C

= 427.5 KJ (student to calculate & match)

Here $(\Delta S)_{ref} = 0$

$$(\Delta S)_{atm} = \frac{Q + W}{T} = \frac{427.5 + W}{293}$$

$$(\Delta S)_{system} = -1.5549 \text{ KJ/K (student to calculate \& match)}$$

$$(\Delta S)_{univ} = (\Delta S)_{sys} + (\Delta S)_{ref} + (\Delta S)_{atm}$$

$$= (S_1 - S_4) + 0 + \frac{427.5 + W}{293}$$

$$= -1.5549 + \frac{427.5 + W}{293} \geq 0$$

or $$\frac{427.5 + W}{293} \geq 1.5549$$

or $$427.5 + W \geq 455.58$$

or $$W \geq 28.08, W_{min} = 28.08 \text{ KJ}.$$

Example 8.5. *How much does the entropy of 1 kg of ice change as it melts into water in each case (i) Heat is supplied reversibly to mixture of ice and water at 0°C (ii) a mixture of ice and water at 0°C is stirred by a paddle wheel. (Latent heat of fusion of water at 0°C = 335 KJ/Kg)*

Solution.

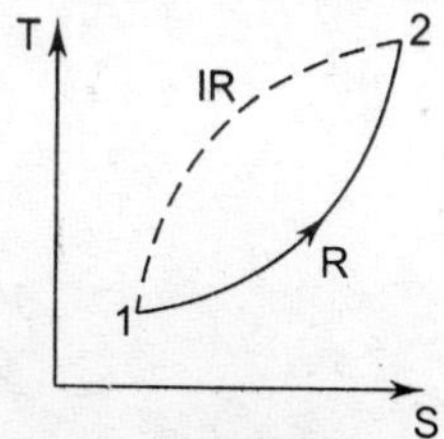

Fig. 8.24

(*i*) $$(\Delta S)_{system} = \frac{Q_{rev}}{T} = \frac{335}{273 + 0} = 1.2271 \text{ KJ/K.} \quad \textbf{Ans.}$$

(*ii*) $(\Delta S)_{system} = 1.2271$ KJ/K

because the irreversible path is replaced by reversible path. As S is a point function there will be same entropy rise to both cases.

Example 8.6. *3 kg of water at 30°C is mixed with 2 kg of water at 80°C adiabatically at a constants pressure of 1 atm. Find the increase in the entropy of the total mass of water due to the mixing (c_p of water = 4.187 KJ/kgk)*

Solution.

Let $$m_1 = 2 \text{ kg}, T_1 = 273 + 80 = 353 \text{ K}, c_1 = 4.187$$

$$m_2 = 3 \text{ kg}, T_2 = 273 + 30 = 303 \text{ K}, c_2 = 4.187$$

using equation (8.20) we get,

$$T_f = \frac{m_1 c_1 T_1 + m_2 c_2 T_2}{m_1 c_1 + m_2 c_2}$$

$$= \frac{2 \times 4.187 \times 353 + 3 \times 4.187 \times 303}{2 \times 4.187 + 3 \times 4.187} = \frac{6762.005}{20.935} = 323\text{K}.$$

From equation (8.21) we get

$$(\Delta S)_{univ} = m_1 c_1 \ln \frac{T_f}{T_1} + m_2 c_2 \ln \frac{T_f}{T_2}$$

$$= 2 \times 4.817 \ln \frac{323}{353} + 3 \times 4.187 \ln \frac{323}{303}$$

$$= -0.744 + 0.803 = 0.059 \text{ KJ/K.} \quad \textbf{Ans.}$$

Example 8.7. *Given $C_V = A + 2BT$ where $A = 0.014$ J/K and $B = 4.2 \times 10^{-4}$ J/K^2, C_V = heat capacity valid from 0°C – 100°C. What is the maximum amount of work that can be transferred to the reversible work source as the system is cooled from 100°C to reservoir temperature (0°C).*

Solution.

Heat removed from system, Q

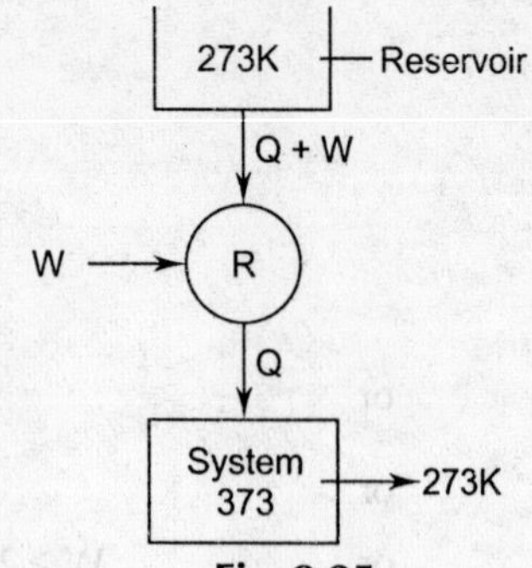

Fig. 8.25

$$Q = \int_{373}^{273} C_V \, dT$$

$$= \int_{373}^{273} (A + 2BT) dT$$

$$= AT \Big|_{373}^{273} + 2B \times \frac{T^2}{2}\Big|_{373}^{273}$$

$$= A\,(273 - 373) + B\,(273^2 - 373^2)$$

$$= -100\,A - 64600\,B$$

$$= -100 \times 0.014 - 64600 \times 4.2 \times 10^{-4}\ \text{J}$$

$$= -1.4 - 27.132 = -28.532\ \text{J}$$

$$(\Delta S)_{res} = \frac{Q + W}{273} = \frac{28.532 + W}{273}$$

$$(\Delta S)_{system} = \int_{373}^{273} C_V \frac{dT}{T} = \int_{373}^{273} (A + 2BT)\frac{dT}{T}$$

$$= A \ln \frac{273}{373} + 2B\,(273 - 373)$$

$$= 0.014 \ln \frac{273}{373} - 200 \times 4.2 \times 10^{-4}$$

$$= -4.37 \times 10^{-3} - 84 \times 10^{-3} = -88.37 \times 10^{-3}\ \text{J/K}$$

$$(\Delta S)_{\text{working fluid in } m/c} = 0$$

$$(\Delta S)_{univ} = (\Delta S)_{system} + (\Delta S)_{res}$$

$$= -88.37 \times 10^{-3} + \frac{28.532 + W}{273} \geq 0$$

or $$\frac{28.532 + W}{273} \geq 88.37 \times 10^{-3}$$

or $$28.532 + W \geq 24.125$$

or $$W \geq -4.407$$

or $$W \leq 4.407$$

∴ $$W_{max} = 4.407. \quad \textbf{Ans.}$$

Chapter 9

Properties of Gases (Real & Ideal)

9.1 IDEAL GAS OR PERFECT GAS

A gas which follows $Pv = RT$ at all pressures and temperatures is called an ideal or perfect gas.

In
$$Pv = RT \quad \text{...(9.1)}$$

P = Pressure (N/m^2), v = specific volume $\left(\frac{m^3}{kg}\right)$.

R = Gas constant which depends on type of gas. (J/kg K)

T = Absolute temperature (K).

If the mass of the gas is m (kg), then it can be included in equation (9.1) by following.

$$v = \frac{V}{m}$$

where V = volume of m kg of gas $\Rightarrow$ (m^3)

(9.1) $\Rightarrow$
$$P\frac{V}{m} = RT$$

or
$$PV = mRT \quad \text{...(9.2)}$$

Equation (9.2) can also be called ideal gas equation of state.

Unit of R:
$$R = \frac{PV}{mT} \Rightarrow \frac{\frac{N}{m^2} \times m^3}{kg\,K} = \frac{N\,m}{kg\,K} = J/kg\,K.$$

Sample calculation for R for air:

At 0°C and at 1.013 bar (at NTP), the volume of 1kg of air is 0.774 m^3.

$$\therefore \quad R = \frac{PV}{mT} = \frac{(1.013 \times 10^5) \times 0.774}{1 \times 273.15} = 287 \text{ Nm/kg K}$$

Multiplying equation (9.1) both sides by M (molecular weight of the gas), we get

$$MPv = MRT \qquad [v \text{ is the volume per unit mass of the gas}]$$

or
$$P(Mv) = (MR)T$$

or
$$P\bar{V} = \bar{R}T$$

where
$$\bar{V} = Mv = \text{molar volume}$$

and
$$\bar{R} = MR$$

As per Avogadro's hypothesis; $\bar{V}$ is same for all gases at NTP and therefore $\bar{R}$ must also be same for all gases. This constant $\bar{R}$ which has the same value for all gases is known as universal gas constant and it does not depend upon the type of gas.

S.I. Unit of $\bar{R}$:

$$\bar{R} = MR = \text{J/kg mole K.}$$

S.I. Value of $\bar{R}$:

$$\bar{V} = 22.4 \text{ m}^3\text{/kg mole}$$

$$\bar{R} = \frac{P\bar{V}}{T} = \frac{1.013 \times 10^5 \times 22.4}{273.15} = 8307 \text{ J/kg mole K} \approx 8314.3 \text{ J/kg mole K}$$

Now the R can be calculated from equation $R = \frac{\bar{R}}{M}$.

$$\therefore \quad R_{O_2} = \frac{8.3143}{32} \text{ KJ/}kg\text{K.}$$

$$= 0.260 \text{ KJ/}kg\text{ K} \approx 0.262 \text{ KJ/}kg\text{ K.}$$

$$R_{air} = \frac{8.3143}{28.96} = 0.287 \text{ KJ/}kg\text{ K.}$$

There are 6.023 × 10^{23} molecules per gram mole of a substance. This is known as Avogadro's number (A).

$$\therefore \quad A = 6.023 \times 10^{26} \text{ molecules / } kg \text{ mol.}$$

The universal gas constant divided by A is called the Boltzman constant and denoted by K.

$$\therefore \quad K = \frac{\bar{R}}{A} = \frac{8314.3}{6.023 \times 10^{26}}$$

$$= 1380 \times 10^{-26} = 13.8 \times 10^{-24} \text{ J/molecule K.}$$

9.2 SPECIFIC HEATS, INTERNAL ENERGY AND ENTHALPY OF AN IDEAL GAS

The ideal gas is having its specific heats constant. It follows the equation $Pv = RT$. But real gas is having different specific heats at different temperatures and pressures. The change is less in case of pressure variation.

We have from 1st law

$$đq = du + đw$$

or

$$Tds = du + Pdv$$

or

$$ds = \frac{du}{T} + \frac{Pdv}{T} \quad \text{...(9.3)}$$

The specific internal energy u is assumed to be a function of T and v, *i.e.*,

$$u = f(T, v)$$

or

$$du = \left(\frac{\partial u}{\partial T}\right)_v dT + \left(\frac{\partial u}{\partial v}\right)_T dv \quad \text{...(9.4)}$$

From equations (9.3) and (9.4) we have

$$ds = \frac{1}{T}\left(\frac{\partial u}{\partial T}\right)_v dT + \frac{1}{T}\left[\left(\frac{\partial u}{\partial v}\right)_T + P\right] dv \quad \text{...(9.5)}$$

Again let, $s = f(T, v)$

$\therefore$

$$ds = \left(\frac{\partial s}{\partial T}\right)_v dT + \left(\frac{\partial s}{\partial v}\right)_T dv \quad \text{...(9.6)}$$

Comparing equations (9.5) and (9.6) we can write

$$\left(\frac{\partial s}{\partial T}\right)_v = \frac{1}{T}\left(\frac{\partial u}{\partial T}\right)_v \quad \text{...(9.7)}$$

and

$$\left(\frac{\partial s}{\partial v}\right)_T = \frac{1}{T}\left[\left(\frac{\partial u}{\partial v}\right)_T + P\right] \quad \text{...(9.8)}$$

Differentiating equation (9.7) w.r.t. v when T is const.

$$\frac{\partial^2 s}{\partial T \partial v} = \frac{1}{T}\frac{\partial^2 u}{\partial T \partial v} \quad \text{...(9.9)}$$

Differentiating equation (9.8) w.r.t. T when v is const.

$$\frac{\partial^2 s}{\partial v \partial T} = \frac{1}{T}\frac{\partial^2 u}{\partial v \partial T} + \frac{1}{T}\left(\frac{\partial P}{\partial T}\right)_v - \frac{1}{T^2}\left(\frac{\partial u}{\partial v}\right)_T - \frac{P}{T^2} \quad \text{...(9.10)}$$

From equations (9.9) and (9.10) we get

$$\frac{1}{T}\frac{\partial^2 u}{\partial T \partial v} = \frac{1}{T}\frac{\partial^2 u}{\partial v \partial T} + \frac{1}{T}\left(\frac{\partial P}{\partial T}\right)_v - \frac{1}{T^2}\left(\frac{\partial u}{\partial v}\right)_T - \frac{P}{T^2}$$

or,

$$\left(\frac{\partial u}{\partial v}\right)_T + P = T\left(\frac{\partial P}{\partial T}\right)_v \quad \text{...(9.11)}$$

We know for an ideal gas

$$Pv = RT$$

or $$v\left(\frac{\partial P}{\partial T}\right)_v = R$$

or $$\left(\frac{\partial P}{\partial T}\right)_v = \frac{R}{v} = \frac{P}{T} \qquad ...(9.12)$$

From equations (9.11) and (9.12) we have

$$\left(\frac{\partial u}{\partial v}\right)_T = 0 \qquad ...(9.13)$$

Thus, u does not change when v changes at constant temperature.

We can also show if $u = f(T, P)$; $\left(\frac{\partial u}{\partial P}\right)_T = 0$. Thus u does not change with P at constant temperature. It means u does not change unless T changes.

So $$u = f(T) \qquad ...(9.14)$$

The equation (9.14) is valid for ideal gas only. Equation (9.14) is known as Joule's law.

If $$u = f(T, v)$$

$$\therefore \quad du = \left(\frac{\partial u}{\partial T}\right)_v dT + \left(\frac{\partial u}{\partial v}\right)_T dv$$

and from equation (9.13) we have $\left(\frac{\partial u}{\partial v}\right)_T = 0$, thus

$$du = \left(\frac{\partial u}{\partial T}\right)_v dT$$

or $$du = c_v dT \qquad ...(9.15)$$

Equation (9.15) is valid for ideal gas and for any process of ideal gas. For other substance equation (9.15) is valid for constant volume process only.

Now $$h = u + Pv = u + RT$$

$$\therefore \quad dh = du + R\,dT$$

or $$\frac{dh}{dT} = \frac{c_v dT}{dT} + R = c_v + R$$

or $$c_p = c_v + R \quad \text{or} \quad c_p - c_v = R \qquad ...(9.16)$$

To note $dh = c_p\, dT$ also holds good for ideal gas even when pressure changes. but it is true for other substance for constant pressure process.

The ratio $\frac{c_p}{c_v}$ is written as γ

$$\therefore \quad \frac{c_p}{c_v} = \gamma \qquad ...(9.17)$$

valid for ideal gas and is used in many computation.

Equations (9.16) and (9.17) give rise to

$$c_p = \frac{\gamma R}{\gamma - 1}, \text{ and } c_v = \frac{R}{\gamma - 1}$$

The S.I. unit of c_p and c_v is KJ/kg K.

Molar or molal specific heats:

$$\overline{c_p} = Mc_p \text{ and } \overline{c_v} = Mc_v.$$

Where $\overline{c_p}$ and $\overline{c_v}$ are molar or molal specific heats at constant pressure and constant volume respectively.

Value of γ:

$$\gamma = \frac{5}{3} \text{ for monoatomic gas}$$

$$= \frac{7}{5} \text{ for diatomic gas}$$

$$= \frac{4}{3} \text{ for polyatomic (more than two atoms) gas.}$$

Thus $$1 \le \gamma \le \frac{4}{3}.$$

So, γ depends on the molecular structure of the gas.

c_p and c_v for ideal gas depend only on γ and R. c_v and c_p are independent of temperature and pressure of the gas.

9.3 ENTROPY CHANGE OF AN IDEAL GAS

From equation (9.3) we have

$$ds = \frac{du}{T} + \frac{P\,dv}{T} \qquad [Pv = RT]$$

$$= \frac{c_v\,dT}{T} + \frac{R\,dv}{v} \qquad \left[\text{or } \frac{P}{T} = \frac{R}{v}\right]$$

Entropy change between state 1 – 2

$$= \int_1^2 ds = \int_1^2 c_v \frac{dT}{T} + \int_1^2 R\frac{dv}{v}$$

or $$s_2 - s_1 = c_v \ln \frac{T_2}{T_1} + R \ln \frac{v_2}{v_1} \qquad ...(9.18)$$

we also have $$T ds = dh - v dp.$$

or $$ds = \frac{dh}{T} - \frac{v dp}{T}$$

$$= c_p \frac{dT}{T} - R\frac{dP}{P} \qquad \left[\frac{v}{T} = \frac{R}{P}\right]$$

∴ Entropy change between two states 1 – 2 can also be written as

$$\int_1^2 ds = \int_1^2 c_p \frac{dT}{T} - \int_1^2 R\frac{dP}{P}$$

or
$$s_2 - s_1 = c_p \ln \frac{T_2}{T_1} - R \ln \frac{P_2}{P_1} \qquad ...(9.19)$$

As $c_p - c_v = R$, equation (9.19) can be written as

$$s_2 - s_1 = c_p \ln \frac{T_2}{T_1} - c_p \ln \frac{P_2}{P_1} + c_v \ln \frac{P_2}{P_1}$$

or
$$s_2 - s_1 = c_p \ln \frac{P_1 T_2}{P_2 T_1} + c_v \ln \frac{P_2}{P_1}$$

or
$$s_2 - s_1 = c_p \ln \frac{v_2}{v_1} + c_v \ln \frac{P_2}{P_1} \qquad ...(9.20)$$

Note: $\left[\frac{P_1 v_1}{T_1} = \frac{P_2 v_2}{T_2} \text{ or } \frac{P_1 T_2}{P_2 T_1} = \frac{v_2}{v_1}\right]$

We can use equations (9.18), (9.19), (9.20) for computing the entropy change between two states of an ideal gas.

9.4 REVERSIBLE ADIABATIC PROCESS

9.4.1 Finding out of Governing Equation for Reversible Adiabatic Process

We know $đq = Tds = du + Pdv = c_v\, dT + Pdv$ for adiabatic process $đq = 0$ *i.e.*, $ds = 0$ thus

$$c_v dT = -Pdv \qquad ...(9.21)$$

Again
$$đq = Tds = dh - vdp = c_p dT - vdP$$

or
$$0 = c_p dT - vdP$$

or
$$c_p dT = vdP \qquad ...(9.22)$$

Dividing equation (9.21) by equation (9.22), we get

$$\frac{c_p\, dT}{c_v\, dT} = -\frac{vdP}{Pdv}$$

or
$$\frac{dP}{P} + \frac{c_p}{c_v}\frac{dv}{v} = 0$$

or
$$\frac{dP}{P} + \gamma \frac{dv}{v} = 0$$

or
$$\int \frac{dP}{P} + \int \gamma \frac{dv}{v} = \ln C \quad \text{where C is constant}$$

$$\ln P + \gamma \ln v = \ln C$$

or $\ln (Pv^{\gamma}) = \ln C$

or $Pv^{\gamma} = C$ Where C = constant ...(9.23)

For process 1 – 2 we can have

$$P_1 v_1^{\gamma} = P_2 v_2^{\gamma}$$

or $$\frac{P_2}{P_1} = \left(\frac{v_1}{v_2}\right)^{\gamma} \quad ...(9.24)$$

For ideal gas $Pv = RT$

or $$\frac{P_1 v_1}{T_1} = \frac{P_2 v_2}{T_2}$$

or $$\frac{P_2}{P_1} = \left(\frac{v_1}{v_2}\right) \cdot \frac{T_2}{T_1}$$

(9.24) ⇒ $$\left(\frac{v_1}{v_2}\right) \cdot \frac{T_2}{T_1} = \left(\frac{v_1}{v_2}\right)^{\gamma}$$

or $$\frac{T_2}{T_1} = \left(\frac{v_1}{v_2}\right)^{\gamma - 1} \quad ...(9.25)$$

Again $$\frac{P_1 v_1}{T_1} = \frac{P_2 v_2}{T_2}$$

or $$\frac{v_1}{v_2} = \left(\frac{P_2}{P_1}\right) \times \left(\frac{T_1}{T_2}\right)$$

Thus (9.25) ⇒ $$\frac{T_2}{T_1} = \left(\frac{P_2}{P_1}\right)^{\gamma - 1} \cdot \left(\frac{T_1}{T_2}\right)^{\gamma - 1}$$

or $$\left(\frac{T_2}{T_1}\right)^{1 + \gamma - 1} = \left(\frac{P_2}{P_1}\right)^{\gamma - 1}$$

or $$\frac{T_2}{T_1} = \left(\frac{P_2}{P_1}\right)^{\frac{\gamma - 1}{\gamma}} \quad ...(9.26)$$

Equations (9.24), (9.25) and (9.26) will give the relation among P, v, and T in a reversible adiabatic process for an ideal gas.

9.4.2 The Internal Energy Change of an Ideal Gas for Reversible Adiabatic Process

We have $Tds = du + Pdv = 0$

or $du = - Pdv$

or $$\int_1^2 du = -\int_1^2 P\,dv = -\int_1^2 \frac{C}{v^{\gamma}}\,dv$$

or $$u_2 - u_1 = -C\left(\frac{v^{-\gamma+1}}{-\gamma+1}\right)\Bigg|_1^2$$

$$= \frac{C}{\gamma-1}\left[v_2^{1-\gamma} - v_1^{1-\gamma}\right]$$

$$= \frac{1}{\gamma-1}\left[Cv_2^{1-\gamma} - Cv_1^{1-\gamma}\right]$$

$$= \frac{1}{\gamma-1}\left[P_2 v_2^{\gamma}\, v_2^{-\gamma+1} - P_1 v_1^{\gamma}\, v_1^{1-\gamma}\right]$$

$$= \frac{1}{\gamma-1}\left[P_2 v_2 - P_1 v_1\right]$$

$$= \frac{R(T_2 - T_1)}{\gamma-1} = \frac{RT_1}{\gamma-1}\left[\frac{T_2}{T_1} - 1\right]$$

$$= \frac{RT_1}{\gamma-1}\left[\left(\frac{P_2}{P_1}\right)^{\frac{\gamma-1}{\gamma}} - 1\right] \qquad ...(9.27)$$

9.4.3 The Enthalpy Change of an Ideal Gas for a Reversible Adiabatic Process

$$T\,ds = dh - v\,dP = 0$$

or $$dh = v\,dP$$

$$\int_1^2 dh = \int_1^2 v\,dP$$

or $$h_2 - h_1 = \int_1^2 \frac{C^{\frac{1}{\gamma}}\,dP}{P^{\frac{1}{\gamma}}}$$

$$\left[P v^{\gamma} = C \text{ or } v = \left(\frac{C}{P}\right)^{\frac{1}{\gamma}}\right]$$

$$= C^{\frac{1}{\gamma}} \int_1^2 P^{-\frac{1}{\gamma}}\,dP = C^{\frac{1}{\gamma}} \left|\frac{P^{-\frac{1}{\gamma}+1}}{-\frac{1}{\gamma}+1}\right|_1^2$$

or $$h_2 - h_1 = C^{\frac{1}{\gamma}}\left[\frac{P_2^{-\frac{1}{\gamma}+1} - P_1^{-\frac{1}{\gamma}+1}}{-\frac{1}{\gamma}+1}\right]$$

$$= \frac{\gamma}{\gamma-1}\left[C^{\frac{1}{\gamma}} P_2^{-\frac{1}{\gamma}+1} - C^{\frac{1}{\gamma}} P_1^{-\frac{1}{\gamma}+1}\right]$$

$$= \frac{\gamma}{\gamma-1}\left[P_2^{\frac{1}{\gamma}} v_2 P_2^{-\frac{1}{\gamma}+1} - P_1^{\frac{1}{\gamma}} v_1 P_1^{-\frac{1}{\gamma}+1}\right]$$

$$= \frac{\gamma}{\gamma-1}[P_2 v_2 - P_1 v_1]$$

$$= \frac{\gamma R T_1}{\gamma-1}\left[\left(\frac{P_2}{P_1}\right)^{\frac{\gamma-1}{\gamma}} - 1\right] \quad \text{...(9.28)}$$

9.4.4 The Workdone by an Ideal Gas in a Reversible Adiabatic Process

We know $\quad đQ = dU + đW = 0$

or $\quad đW = -dU$

i.e., work is done at the expense of the internal energy. If m is the mass of the gas.

Then $\therefore \quad W_{1-2} = U_1 - U_2 = m(u_1 - u_2).$

$$= \frac{m(p_1 v_1 - p_2 v_2)}{\gamma-1} = \frac{mR(T_1 - T_2)}{\gamma-1} = \frac{mRT_1}{\gamma-1}\left[1-\left(\frac{P_2}{P_1}\right)^{\frac{\gamma-1}{\gamma}}\right] \quad \text{...(9.29)}$$

9.4.5 Reversible Adiabatic Steady Flow Process

From equation (5.9) in Chapter 5, we have

$$q + \left(h + \frac{\overline{V}^2}{2} + gz\right)_{in} = \left(h + \frac{\overline{V}^2}{2} + gz\right)_{out} + w \qquad \text{[in = state 1, out = state 2]}$$

For process 1 – 2

or $$q_{1-2} + \left(h_1 + \frac{\overline{V}_1^2}{2} + gz_1\right) = h_2 + \frac{\overline{V}_2^2}{2} + gz_2 + w_{1-2}$$

For adiabatic $q = 0$

or, $$0 + h_1 + \frac{\overline{V}_1^2}{2} + gz_1 = h_2 + \frac{\overline{V}_2^2}{2} + gz_2 + w_{1-2}$$

or, $$w_{1-2} + \frac{\overline{V}_2^2 - \overline{V}_1^2}{2} + g(z_2 - z_1) = h_1 - h_2$$

$$= c_p (T_1 - T_2) = \frac{\gamma R}{\gamma - 1}(T_1 - T_2), \qquad \text{[Putting value of } c_p]$$

$$= \frac{\gamma}{\gamma - 1}(P_1 v_1 - P_2 v_2) = \frac{\gamma}{\gamma - 1} P_1 v_1 \left[1 - \left(\frac{P_2}{P_1}\right)^{\frac{\gamma - 1}{\gamma}}\right]$$

If kinetic and potential energy neglected, then

$$w_{1-2} = \frac{\gamma P_1 v_1}{\gamma - 1}\left[1 - \left(\frac{P_2}{P_1}\right)^{\frac{\gamma - 1}{\gamma}}\right] \qquad ...(9.30)$$

9.4.6 Reversible Isothermal Process

If ideal gas changes its state from 1 to 2 isothermally, the workdone is given by

$$\int_1^2 dW = \int_{V_1}^{V_2} P\,dV = \int_{V_1}^{V_2} \frac{mRT}{V}\,dV \qquad [PV = mRT]$$

$$= mRT \ln \frac{V_2}{V_1} \qquad \left[P = \frac{mRT}{V}\right]$$

or
$$W_{1-2} = mRT \ln \frac{V_2}{V_1} = mRT \ln \frac{P_1}{P_2} \qquad ...(9.31)$$

The heat transfer Q_{1-2} is given by 1st law,

$$Q_{1-2} = U_2 - U_1 + W_{1-2}$$

$$= 0 + W_{1-2}$$

$$= mRT \ln \frac{V_2}{V_1} = mRT \ln \frac{P_1}{P_2} \qquad ...(9.32)$$

or
$$Q_{1-2} = T(S_2 - S_1)$$

$[S_2 - S_1$ = entropy change (not specific entropy change)].

9.4.7 Polytropic Process

In practice Pv^n = constant is an equation of polytropic process. Here n is the index of a process. This equation is an equation of many processes depending upon the value of n. γ is a property of the gas but n is not.

$$Pv^n = \text{constant}$$

For two states 1 – 2, $P_1 v_1^n = P_2 v_2^n$

or
$$\ln P_1 + n \ln v_1 = \ln P_2 + n \ln v_2$$

or
$$n = \frac{\ln P_1 - \ln P_2}{\ln v_2 - \ln v_1} \qquad ...(9.33)$$

Thus n can be found out from P_1, P_2, v_1 and v_2 values.

Again $\qquad P_1 v_1^n = P_2 v_2^n$

$$P_1 v_1 v_1^{n-1} = P_2 v_2 v_2^{n-1}$$

or $$RT_1 v_1^{n-1} = RT_2 v_2^{n-1}$$

or $$\frac{T_2}{T_1} = \left(\frac{v_1}{v_2}\right)^{n-1} \quad ...(9.34)$$

or $$\frac{T_2}{T_1} = \left(\frac{P_2}{P_1}\right)^{\frac{n-1}{n}} \quad ...(9.35)$$

9.4.7.1 *Entropy Change in a Polytropic Process*

In a reversible adiabatic process $dS = 0$. But in a polytropic process the entropy changes. From equation (9.18) we can write

$$s_2 - s_1 = c_v \ln \frac{T_2}{T_1} + R \ln \frac{v_2}{v_1}$$

$$= \frac{R}{\gamma - 1} \ln \frac{T_2}{T_1} + \frac{R}{n-1} \ln \frac{T_1}{T_2}$$

$$= \frac{(n-\gamma)R}{(\gamma-1)(n-1)} \ln \frac{T_2}{T_1} \quad ...(9.36)$$

Similarly $$s_2 - s_1 = \left(\frac{n-\gamma}{(\gamma-1)(n-1)} R \ln \frac{P_2}{P_1}\right) \quad ...(9.37)$$

and $$s_2 - s_1 = \frac{n-\gamma}{(\gamma-1)(n-1)} R \ln \frac{v_1}{v_2} \quad ...(9.38)$$

From equations (9.36), (9.37) and (9.38), if $n = \gamma$ is put then $s_2 - s_1 = 0$ *i.e.*, the specific entropy change becomes zero.

For $n \le \gamma, s_1 > s_2$, *i.e.*, specific entropy of the gas decreases.

and for $n > \gamma, s_2 > s_1$, *i.e.*, specific entropy of the gas increases.

If the gas is cooled, the entropy decrease is possible, and if reversible heat transfer takes place, the increase of entropy is possible.

9.4.7.2 *Heat and Work in a Polytropic Process*

From 1st law

$$q_{1-2} = u_2 - u_1 + w_{1-2}$$

or $$q_{1-2} - w_{1-2} = u_2 - u_1 = c_v (T_2 - T_1)$$

$$= \frac{R}{\gamma-1}(T_2 - T_1) = \frac{P_2 v_2 - P_1 v_1}{\gamma - 1}$$

$$= \frac{P_1 v_1}{\gamma-1}\left[\left(\frac{P_2}{P_1}\right)^{\frac{n-1}{n}} - 1\right] = \frac{P_1 v_1}{\gamma-1}\left[\left(\frac{v_1}{v_2}\right)^{n-1} - 1\right] \quad ...(9.39)$$

For steady flow process, we have

$$q_{1-2} - w_{1-2} = h_2 - h_1 \quad \text{[K.E. \& P.E. neglected]}$$
$$= c_p (T_2 - T_1)$$
$$= \frac{\gamma R}{\gamma - 1}(T_2 - T_1)$$
$$= \frac{\gamma}{\gamma - 1}(P_2 v_2 - P_1 v_1) \quad ...(9.40)$$

For a polytropic process (K.E. and P.E. neglected)

$$q_{1-2} - w_{1-2} = \frac{\gamma P_1 v_1}{\gamma - 1}\left[\left(\frac{P_2}{P_1}\right)^{\frac{n-1}{n}} - 1\right]$$

$$= \frac{\gamma P_1 v_1}{\gamma - 1}\left[\left(\frac{v_1}{v_2}\right)^{n-1} - 1\right] \quad ...(9.41)$$

Equations (9.39) and (9.41) are used to determine heat and work quantities for a closed and a steady flow system respectively.

9.4.7.3 *Polytropic Specific Heat*

For $\quad Pv^n = \text{constant}$

$$\int_1^2 P\,dv = \int_1^2 \frac{P_1 v_1^n}{v^n}\,dv = \frac{P_1 v_1}{n-1}\left[1 - \left(\frac{v_1}{v_2}\right)^{n-1}\right]$$

$$= \frac{P_1 v_1}{n-1}\left[1 - \left(\frac{P_2}{P_1}\right)^{\frac{n-1}{n}}\right] \quad ...(9.42)$$

Similarly,

$$-\int_1^2 v\,dP = \frac{nP_1 v_1}{n-1}\left[1 - \left(\frac{v_1}{v_2}\right)^{n-1}\right]$$

$$= \frac{nP_1 v_1}{n-1}\left[1 - \left(\frac{P_2}{P_1}\right)^{\frac{n-1}{n}}\right] \quad ...(9.43)$$

Now $\quad T ds = du + P dv$

or
$$\int_1^2 T\,ds = \int_1^2 du + \int_1^2 P\,dv$$

$$= u_2 - u_1 + \int_1^2 P\,dv$$

Substituting from equations (9.39) and (9.42) we have

$$\int_1^2 \mathrm{T}\,ds = \frac{\mathrm{P}_1 v_1}{\gamma-1}\left\{\left(\frac{\mathrm{P}_2}{\mathrm{P}_1}\right)^{\frac{n-1}{n}} - 1\right\} + \frac{\mathrm{P}_1 v_1}{n-1}\left\{1-\left(\frac{\mathrm{P}_2}{\mathrm{P}_1}\right)^{\frac{n-1}{n}}\right\}$$

$$= \frac{(\gamma-n)\mathrm{P}_1 v_1}{(\gamma-1)(n-1)}\left[1-\left(\frac{\mathrm{P}_2}{\mathrm{P}_1}\right)^{\frac{n-1}{n}}\right]$$

$$= \frac{(\gamma-n)\mathrm{P}_1 v_1}{(\gamma-1)(n-1)}\left[1-\left(\frac{v_1}{v_2}\right)^{n-1}\right]$$

$$= \frac{(\gamma-n)}{(\gamma-1)(n-1)}\left[\mathrm{P}_1 v_1 - \frac{\mathrm{P}_1 v_1^n}{v_2^{n-1}}\right]$$

$$= \frac{(\gamma-n)}{(\gamma-1)(n-1)}\left[\mathrm{P}_1 v_1 - \frac{\mathrm{P}_2 v_2^n}{v_2^{n-1}}\right]$$

$$= \frac{(\gamma-n)}{(\gamma-1)(n-1)}[\mathrm{P}_1 v_1 - \mathrm{P}_2 v_2]$$

$$= \frac{(\gamma-n)\mathrm{R}}{(\gamma-1)(n-1)}(\mathrm{T}_1-\mathrm{T}_2) \qquad ...(9.44)$$

As $c_v = \dfrac{\mathrm{R}}{\gamma-1}$, and let $\Delta\mathrm{T} = \mathrm{T}_2 - \mathrm{T}_1$

so $$\int_1^2 \mathrm{T}\,ds = \frac{c_v(\gamma-n)\Delta\mathrm{T}}{1-n}$$

or $$(q_{1-2})_{rev} = \frac{c_v(\gamma-n)\Delta\mathrm{T}}{1-n}$$

$$= c_n\,\Delta\mathrm{T}$$

where $c_n = \dfrac{c_v(\gamma-n)}{1-n}$, and c_n is called the polytropic specific heat.

For $n > \gamma \Rightarrow (q_{1-2})_{rev} > 0$ *i.e.*, positive heat transfer and thus gain in entropy.

For $n < \gamma \Rightarrow (q_{1-2})_{rev} < 0$ *i.e.*, negative heat transfer and thus loss in entropy.

Generally heat and work both are involved in a polytropic process. To calculate heat transfer during the polytropic process, it is first to evaluate the work via either $\int \mathrm{P}\,dv$ or $-\int v\,dp$ for closed or open steady flow process respectively. By applying 1st law, we will then get heat transfer.

9.5 P – v AND T – s DIAGRAMS FOR $Pv^n = C$

Now $$Pv^n = C$$

or $$v^n\,dP + Pnv^{n-1}\,dv = 0$$

or $$\frac{dP}{dv} = -\frac{nP}{v}$$

or, $$\text{Slope} = -\frac{nP}{v}.$$

Thus slope depends on P and v for a particular n, in negative sense.

Process	Value of n	Equation of process
Isobaric Process	$n = 0$	P = Const.
Isothermal Process	$n = 1$	T = Const.
Isentropic Process	$n = \gamma$	S = Const.
Isometric or isochoric Process	$n = \infty$	v = Const.

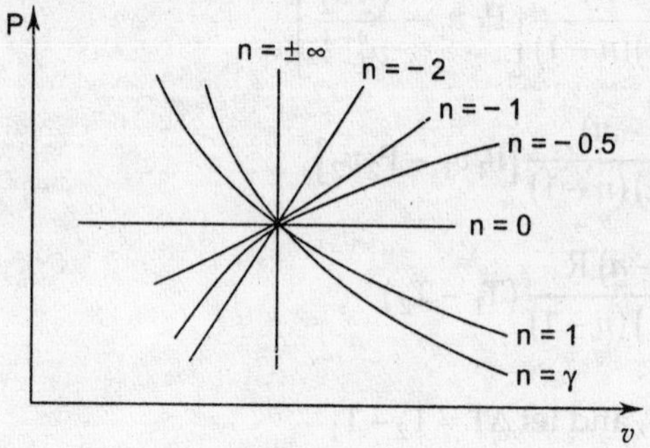

Fig. 9.1

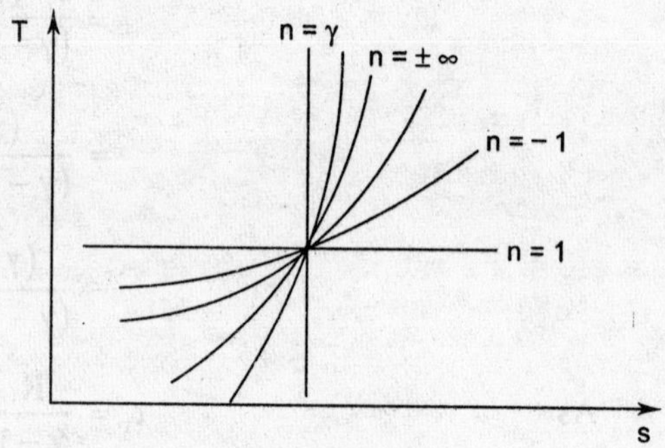

Fig. 9.2

Check for $$n = \pm\infty$$

$$P_1 v_1^n = P_2 v_2^n$$

or $$\frac{v_1}{v_2} = \left(\frac{P_2}{P_1}\right)^{\frac{1}{n}} = \left(\frac{P_2}{P_1}\right)^{\frac{1}{\pm\infty}} = 1$$

$\therefore$ $$v_1 = v_2$$

Similarly $v_2 = v_3$ and so on

$\therefore$ v = Constant for $n = \pm\infty$

9.6 EQUATIONS OF STATE

The ideal gas equation Pv = RT is a consequence of Clerk Maxwells, kinetic theory of gases. The following assumptions are made in kinetic theory.

(*a*) The molecules are perfectly elastic and rigid. Thus there is no loss of momentum during collisions of the molecules with the walls of the container.

(*b*) The volume occupied by the molecules themselves is negligible as compared to the volume of the gas.

(c) There are no inter molecular forces of attraction or repulsion.

When pressure is very small or temperature is very high the real gas obeys very closely the ideal gas equation.

But as pressure increases the point (c) is no longer valid. The intermolecular forces of attraction or repulsion increases, and also the point (b) is invalid. The volume of molecules are appreciable compared to the total gas volume.

For real gas the following equations are valid.

(1) $$\left(P + \frac{a}{v^2}\right)(v - b) = RT \rightarrow \text{van der Waal's equation}$$

where $\left(\frac{a}{v^2}\right)$ takes into account for intermolecular forces of attraction and b is called covolume taking account for volumes of molecules.

(2) $$P = \frac{RT}{v-b} - \frac{a}{Tv^2} \rightarrow \text{Berthelot's equation}$$

(3) $$P = \frac{RT}{v-b} e^{\frac{-a}{RTv}} \rightarrow \text{Dieterici's equation}$$

(4) $$P = \frac{RT}{v-b} - \frac{a}{v(v+b)\sqrt{T}} \rightarrow \text{Redlich-Kwong's equation}$$

(5) $$P = \frac{-RT}{2b} e^{\frac{-a}{RTv}} \ln\left(\frac{v-2b}{v}\right) \rightarrow \text{Saha-Bose equation}$$

(6) $$P = \frac{RT(1-e)}{v^2}(v+B) - \frac{A}{v^2} \rightarrow \text{Beatlie-Bridgeman equation}$$

where $$A = A_0\left(1 - \frac{a}{v}\right), B = B_0\left(1 - \frac{b}{v}\right), e = \frac{C}{vT^3}.$$

A_0, B_0, a, b, c are constants.

Table 9.1. Van der Waals constants

Gas	a $(Nm^4/(K\ mol)^2$	b $m^3/(K\ mol)$
Air	137,052	0.0366
NH_3	426,295	0.0373
CO_2	368,127	0.0428
Fr_{12}	1,082,470	0.0998
He	3,440	0.0232
H_2	24,800	0.0266
N_2	137,450	0.0378
O_2	139,044	0.0317
H_2O (vapour)	561,753	0.0317

9.7 LAW OF CORRESPONDING STATES

We have $v_{\text{ideal}} = \dfrac{RT}{P}$

$$\frac{v_{real}}{v_{ideal}} = \text{compressibility factor} = Z$$

or $$Z = \frac{v}{\frac{RT}{P}} = \frac{Pv}{RT} \qquad [\text{Let } v_{\text{real}} = v]$$

or $$Pv = ZRT. \qquad ...(9.45)$$

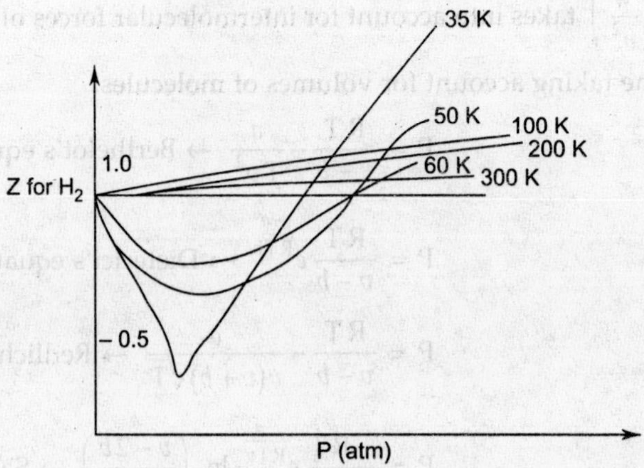

Fig. 9.3

Fig. 9.3 shows a plot of Z against P and T, which is also called compressibility factor chart.

Separate charts are needed for every gas. However, one single generalized chart can be developed based on law of corresponding states where we express pressure, temperature and specific volume in dimensionless forms

$$P_r = \frac{P}{P_c},\ T_r = \frac{T}{T_c},\ v_r = \frac{v}{v_c}$$

where P_r, T_r and v_r are known as the reduced pressure, reduced temperature and reduced specific volume respectively, and P_c, T_c and v_c are the critical properties of substance.

Now the law of corresponding states may be stated as follows :

"If two or more substances have the same reduced pressure P_r, and reduced temperature T_r, then they will have the same reduced volume v_r"

Mathematically $\quad Z = f(P_r, T_r) \qquad ...(9.46)$

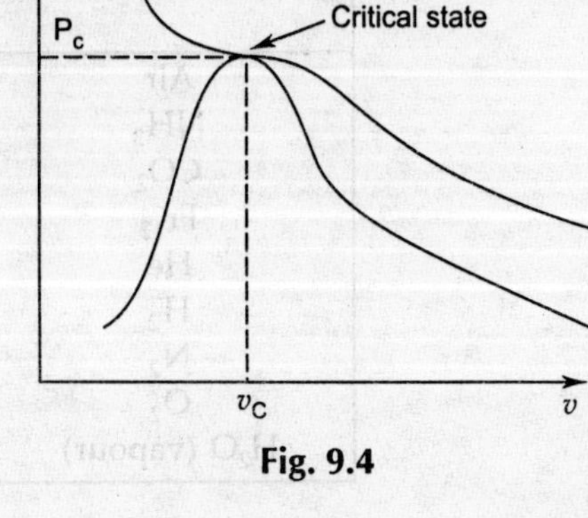

Fig. 9.4

$$v_r = \frac{v}{v_c} = \frac{\frac{ZRT}{P}}{\frac{Z_c RT_c}{P_c}} \qquad [Pv = ZRT]$$

$$= \frac{ZRTP_c}{Z_c RT_c P} = \frac{Z}{Z_c} \cdot \frac{T_r}{P_r} \qquad ...(9.47)$$

where $Z_c = \dfrac{P_c v_c}{RT_c}$ = critical compressibility factor.

When T_r is ploted as a function of reduced pressure and Z, a single plot called the generalized compressibility chart is obtained for a great variety of substances.

We have $\left(P + \dfrac{a}{v^2}\right)(v - b) = RT$

or $Pv^3 - (Pb + RT)v^2 + av - ab = 0$

or
$$\left(\frac{\partial P}{\partial v}\right)_{T=T_c} = \frac{-RT_c}{(v_c - b)^2} + \frac{2a}{v_c^3} = 0 \qquad ...(9.48)$$

$$\left(\frac{\partial^2 P}{\partial v^2}\right)_{T=T_c} = \frac{2R.T_c}{(v_c - b)^3} - \frac{6a}{v_c^4} = 0 \qquad ...(9.49)$$

From equations (9.48) and (9.49), we get

$$b = \frac{v_c}{3} \text{ and } R = \frac{8a}{9T_c v_c}$$

Substituting the values of R and b in van der Waals equation

we get
$$\left(P_c + \frac{a}{v_c^2}\right)(v_c - b) = RT_c$$

or
$$\left(P_c + \frac{a}{v_c^2}\right)\left(v_c - \frac{v_c}{3}\right) = \frac{8a}{9T_c v_c}.T_c$$

or
$$a = 3P_c v_c^2.$$

Therefore the value of R becomes

$$R = \frac{8}{3}\frac{P_c v_c}{T_c}.$$

Thus we have found out a, b and R in terms of critical properties. So van der Waals equation of state becomes

$$\left(P + \frac{3P_c v_c^2}{v^2}\right)\left(v - \frac{v_c}{3}\right) = \frac{8P_c v_c}{3T_c}.T$$

or $$\left(\frac{P}{P_c}+\frac{3v_c^2}{v^2}\right)\left(\frac{v}{v_c}-\frac{1}{3}\right)=\frac{8T}{3T_c}$$

or $$\left(P_r+\frac{3}{v_r^2}\right)(3v_r-1)=8T_r \qquad ...(9.50)$$

Equation (9.50) is an expression of the law of corresponding state.

We have from above last equation (9.50)

$$P_r v_r=\frac{8T_r v_r}{3v_r-1}-\frac{3}{v_r} \qquad ...(9.51)$$

For minima isotherms we have

$$\frac{\partial}{\partial P_r}(P_r v_r)=\frac{\partial}{\partial P_r}\left[\frac{8T_r v_r}{3v_r-1}-\frac{3}{v_r}\right]_{T_r}=0$$

or $$\frac{\partial}{\partial v_r}\left[\frac{8T_r v_r}{3v_r-1}-\frac{3}{v_r}\right]_{T_r}\left[\frac{\partial v_r}{\partial P_r}\right]_{T_r}=0$$

Since $$\left[\frac{\partial v_r}{\partial P_r}\right]_{T_r}\neq 0$$

$$\frac{\partial}{\partial v_r}\left[\frac{8T_r v_r}{3v_r-1}-\frac{3}{v_r}\right]_{T_r}=0$$

or $$\frac{8T_r}{(3v_r-1)^2}=\frac{3}{v_r^2}$$

or $$\frac{3(3v_r-1)^2}{v_r^2}=8T_r=\left[P_r+\frac{3}{v_r^2}\right][3v_r-1]$$

Simplifying

$\therefore$ $$(P_r v_r)^2-9P_r v_r+6P_r=0 \Rightarrow \text{a parabola}$$

Passing through minima of the isotherms.

When $$P_r=0$$

$$P_r v_r = 0 \text{ and } 9.$$

$$P_r=\frac{9(P_r v_r)-(P_r v_r)^2}{6}$$

$$\frac{dP_r}{d(P_r v_r)}=9-2(P_r v_r)=0$$

$\therefore$ $$P_r v_r=4.5$$

$$P_r=\frac{9\times 4.5-4.5^2}{6}=3.375$$

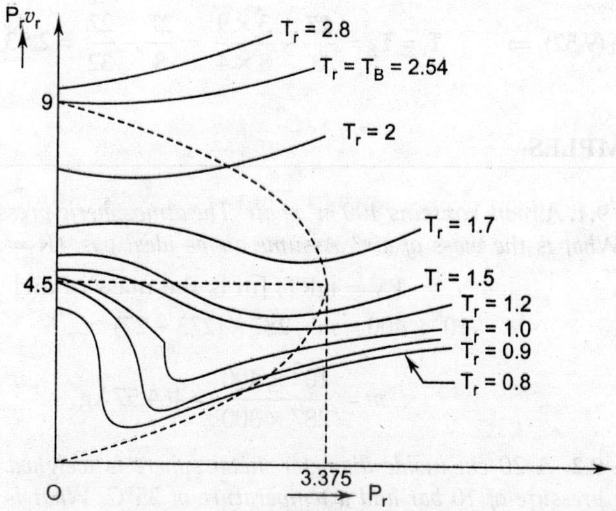

Fig. 9.5

The parabola has the vertex at $P_r\, v_r = 4.5$ and $P_r = 3.375$ and the parabola intersects the ordinate at 0 and 9.

Boyle Temperature (T_B) : Every isotherm has a minimum upto the marked T_B. The T_B isotherm has a starting equation $P_r v_r$ = constants which obeys Boyle's law at moderate pressure. The corresponding temperature is called Boyle temperature.

We have $$\left(P_r\, v_r + \frac{3}{v_r}\right)(3v_r - 1) = 8\, T_r\, v_r$$

or $$\left(9 + \frac{3}{v_r}\right)(3v_r - 1) = 8\, T_r\, v_r$$

or $$3\,(3v_r + 1)\,(3v_r - 1) = 8\, T_r\, v_r^2$$

or $$3\,(9v_r^2 - 1) = 8\, T_r\, v_r^2$$

or $$T_r = \frac{27}{8} - \frac{3}{8v_r^2} \quad \text{for } P_r v_r = 9 \qquad \text{...(9.52)}$$

Again for $$P_r\, v_r = 0$$

$$\frac{3}{v_r}(3v_r - 1) = 8\, T_r\, v_r$$

or $$\frac{9}{8v_r} - \frac{3}{8v_r^2} = T_r$$

Differentiating $$-\frac{9}{8v_r^2} + \frac{6}{8v_r^3} = 0 \text{ for } T_r = T_B.$$

or $$\frac{9}{8v_r^2} = \frac{6}{8v_r^3}$$

or $$v_r = \frac{2}{3}$$

$\therefore$ From (9.52) $\Rightarrow$ $T_r = T_B = \dfrac{27}{8} - \dfrac{3\times 9}{8\times 4} = \dfrac{27}{8} - \dfrac{27}{32} = 2.53$, for $P_r v_r = 9$.

SOLVED EXAMPLES

Example 9.1. *A room contains 400 m³ of air. The atmospheric pressure is 1 bar, and temperature is 27°C. What is the mass of air? Assume air as ideal gas. (R = 287).*

Solution. $PV = mRT$, [m is the mass of air] $1 \text{ bar} = 10^5 \text{ N}/m^2$.

or $$10^5 \times 400 = m \times 287 \times (273 + 27)$$

or $$m = \frac{10^5 \times 400}{287 \times 300} = 464.57 \text{ kg}.$$

Example 9.2. *A 20 cm inside diameter metal sphere is weighed to contain 2.54 gm of a gas filled at a pressure of 10 bar and a temperature of 25°C. What is the likely gas in the sphere? ($\bar{R}$ = 8314.4 units)*

Solution.

$$V = \frac{\pi D^3}{16} = \frac{\pi(0.2)^3}{16} = 0.00157 \ m^3$$

$$PV = mRT$$

or $$10 \times 10^5 \times 0.00157 = 2.54 \times 10^{-3} \times R \times 298$$

$\therefore$ $$R = 2074.2$$

$\therefore$ $$MR = \bar{R} = 8314.4 \text{ [M = Molecular weight]}$$

or $$M = \frac{8314.4}{2074.2} = 4.008$$

Ans : The gas is Helium (He).

Example 9.3. *The density of a gas at NTP is 1.4 kg/m³. Calculate the gas constant and molecular weight of the gas. Find the specific volume of the gas at 14 bar and 40°C.*

Solution. $$Pv = RT$$

$$R = \frac{Pv}{T} = \frac{P}{\rho T} = \frac{1.013 \times 10^5}{1.4 \times (273 + 0^\circ)} \quad \text{[NTP} \Rightarrow 1.013 \text{ bar \& } 0^\circ\text{C]}$$

$$= 265.04 \text{ Nm}/kg\text{ K. (= Gas constant).}$$ **Ans.**

Universal Gas constant $$\bar{R} = MR$$

or $$M = \frac{\bar{R}}{R} = \frac{8314.3}{265.04} = 31.37 \text{ (= Molecular weight of the gas)}$$ **Ans.**

Specific volume $$v = \frac{RT}{P} = \frac{265.04 \times (273 + 40)}{14 \times 10^5}$$

$$= 0.0592 \, m^3/kg.$$ **Ans.**

Example 9.4. *A spherical vessel of 2 m in diameter containing air at 45°C is evacuated till the vacuum inside the vessel becomes 73.5 cm of Hg. The temperature inside the vessel remains constant. Find (a) the mass pumped out of vessel (b) If the tank is then cooled to 4°C, what is the pressure inside the tank in Cm of Hg. Barometer reads 76 Cm of Hg?*

Solution. V = volume of the tank

$$= \frac{4}{3}\pi r^3$$

$$= \frac{4}{3}\pi\left(\frac{2}{2}\right)^3 = 4.19\ m^3$$

Absolute pressure in the tank $= P_2 = 76 - 73.5 = 2.5$ cm of Hg.

(*a*) Mass pumped out = Initial mass – final mass

$$= \frac{P_1 V_1}{R T_1} - \frac{P_2 V_2}{R T_2}$$

$$= \frac{V_1}{R T_1}(P_1 - P_2) \qquad [T_1 = T_2 \;\&\; V_1 = V_2]$$

$$= \frac{4.19\,(76 - 2.5) \times 10^{-2} \times 13.6 \times 1000 \times 9.81}{287 \times (273 + 45)}$$

$= 4.5\,kg.$ **Ans.**

(*b*) The vessel is cooled at constant volume, so

$$\frac{P_2}{T_2} = \frac{P_3}{T_3}$$

or

$$P_3 = \frac{P_2}{T_2} \times T_3 = 2.5 \times \frac{(4 + 273)}{(273 + 45)} \text{ cm of Hg}$$

$= 2.177$ cm of Hg. **Ans.**

Example 9.5. *A balloon is considered as a sphere of 20 m in diameter. The surrounding air is at 15°C and barometer reads 75 cm of Hg. What is the load the balloon lifts if it is filled with H_2 at 27°C and atmospheric pressure?*

Solution. V(volume of balloon) $= \frac{4}{3}\pi r^3 = \frac{4}{3}\pi\left(\frac{20}{2}\right)^3 = 4188.8\ m^3$

$$\text{R for } H_2 = \frac{\overline{R}}{M_{H_2}} = \frac{8314.3}{2} = 4157.15 \text{ Nm}/kg\text{ K}$$

Pressure of H_2 = atmospheric pressure

$$= \frac{75}{76} \times 1.013 \times 10^5 \text{ N}/m^2$$

$$= 10^5 \text{ N}/m^2$$

$$\therefore \quad m_{H_2} = \left(\frac{PV}{RT}\right)_{H_2} = \frac{10^5 \times 4188.8}{4157.15 \times (273 + 27)}$$

$$= 335.87\ kg$$

volume of air displaced by balloon = volume of balloon

$$= 4188.8\ m^3$$

Mass of air displaced by balloon

$$m_{air} = \left(\frac{PV}{RT}\right)_{air} = \frac{\frac{75}{76} \times 1.013 \times 10^5 \times 4188.8}{287 \times (273 + 15)}$$

$$= 5066\ kg.$$

$$\therefore \quad \text{Buoyant force} = (5066 - 335.87)\ kg \times 9.81$$

$$= 46402.5\ \text{N} = 46.4025\ \text{KN}. \quad \textbf{Ans.}$$

Example 9.6. *One kg of an ideal gas is heated from 20°C to 100°C. Taking R = 270 Nm/kg K and $\gamma = 1.3$ for the gas, find (i) change in internal energy (ii) change in enthalpy (iii) molar specific heats.*

Solution.

(*i*) change in internal energy = dU

$$= m\,c_v (T_2 - T_1)$$

$$= 1 \times \frac{R}{\gamma - 1} (T_2 - T_1)$$

$$= \frac{270}{0.3} (100 - 20) = 72000\ \text{J} = 72\ \text{KJ}.$$

(*ii*) Change in enthalpy = dH

$$dH = m\,c_p (T_2 - T_1)$$

$$= 1 \times \gamma c_v (T_2 - T_1) = 1.3 \times \frac{270}{0.3} \times 80$$

$$= 93.6\ \text{KJ}.$$

(*iii*) Molar specific heats = $\overline{c_p}$

$$\overline{c_p} = Mc_p = \frac{M \times \gamma R}{\gamma - 1} = (MR)\frac{\gamma}{\gamma - 1} = \frac{\overline{R}\gamma}{\gamma - 1}$$

$$= \frac{8314 \times 1.3}{0.3} = 3.6 \times 10^4\ \text{J/mole K}.$$

$$= 36\ \text{KJ/mole K}$$

$$\overline{c_v} = \frac{\overline{c_p}}{\gamma} = \frac{36}{1.3} = 27.69\ \text{KJ/mole K}.$$

Example 9.7. *A 0.6 m^3 capacity vessel contains O_2 at 12.6 bar. It is stirred by a paddle until its pressure becomes 21 bar. Find (i) dU (ii) work input (iii) heat transfer (iv) change in specific entropy.* *[Take c_v = 0.657 KJ/kg K, c_p = 0.904 KJ/kg K]*

Solution.

$$c_p - c_v = \text{R}$$

or $$0.904 - 0.657 = \text{R}$$

or $$\text{R} = 0.247 \text{ KJ/kg k.}$$

$$d\text{U} = mc_v\,(\text{T}_2 - \text{T}_1) = c_v\,(m\text{T}_2 - m\text{T}_1)$$

$$= c_v\left(\frac{\text{P}_2\,\text{V}_2}{\text{R}} - \frac{\text{P}_1\,\text{V}_1}{\text{R}}\right) \qquad \left[\frac{\text{P}_2\,\text{V}_2}{\text{T}_2} = m\text{R}\right]$$

$$= c_v\,\text{V}\left(\frac{\text{P}_2 - \text{P}_1}{\text{R}}\right) \qquad [\text{V}_1 = \text{V}_2 = \text{V}]$$

$$= \left(\frac{657 \times 0.6}{247}\right)\left(\frac{21 \times 10^5 - 12.6 \times 10^5}{10^3}\right)\text{KJ}$$

$= 1340.6$ KJ. **Ans(*i*).**

From 1st law of thermodynamics

$$\text{Q}_{1-2} = \text{U}_2 - \text{U}_1 + \text{W}_{1-2} \qquad [\text{Heat transfer} = 0]$$ **Ans(*ii*).**

$$0 = 1340.6 + \text{W}_{1-2}$$

$\therefore$ $\text{W}_{1-2} = -1340.6$ KJ. (work input) **Ans(*iii*).**

Change in specific entropy

$$ds = c_v \ln\frac{\text{T}_2}{\text{T}_1} = c_v \ln\left(\frac{\text{P}_2\,\text{V}_2}{\text{P}_1\,\text{V}_1}\right)$$

$$= c_v \ln\left(\frac{\text{P}_2}{\text{P}_1}\right) = 0.657 \times \ln\frac{21}{12.6} \qquad [\text{V}_1 = \text{V}_2]$$

$= 1.095$ KJ/*kg* K. **Ans(*iv*).**

Chapter **10**

Air Standard Cycles

10.1 INTRODUCTION

Practical gas power cycles are complex. We simplify our analysis by making the following approximations which are known as air standard assumptions.

(*a*) The working fluid is an ideal gas.
(*b*) The specific heats (c_p & c_v) are constant.
(*c*) The cyclic process is reversible.
(*d*) There is no dissociation at the end of the heating process.
(*e*) Circulating mass of the working fluid is constant in the cycle.
(*f*) Addition of heat to the cycle is done by an external source.
(*g*) Rejection of heat by the cycle is done to an external sink.
(*h*) There is no change in K.E. & P.E.

10.2 INTERNAL COMBUSTION ENGINE TERMINOLOGY

In internal combustion engine, the combustion takes place internally.

10.2.1 Bore

The inner diameter of the cylinder is called *bore.*

10.2.2 Top Dead Centre (T.D.C.)

The extreme position of the piston at the top of the cylinder (head side) is known as top dead centre (TDC). In case of horizontal engines this is known as the outer dead centre (ODC).

At the dead centres the piston reverses the direction of its motion and its velocity therefore equals to zero at these dead centres.

10.2.3 Bottom Dead Centre (B.D.C.)

The extreme position of the piston at the bottom of the cylinder is known as bottom dead centre (B.D.C.)

10.2.4 Stroke

The distance between the extreme positions of the piston *i.e.*, the distance between two dead centres is called the stroke length. Stroke length is equal to double the crank radius.

10.2.5 Swept Volume

The volume swept through by the piston in moving between top dead centre and bottom dead centre is called the swept volume and is denoted by V_s.

$$\therefore \quad V_s = \frac{\pi}{4} d^2 \times l$$

where d = cylinder bore and l = stroke length.

10.2.6 Clearance Volume

The space above the piston TDC is known as combustion chamber space and its volume is called the clearance volume and is denoted by V_c.

10.2.7 Volume of the Cylinder

It is the sum of the clearance volume and swept volume.

$$\therefore \quad \text{Cylinder volume} = V = V_c + V_s.$$

10.2.8 Compression Ratio

The ratio of the total volume of the cylinder to that of the clearance volume is called the compression ratio (r).

$$\therefore \quad r = \frac{V}{V_c} = \frac{V_c + V_s}{V_c} = 1 + \frac{V_s}{V_c}$$

compression ratio varies from 5 to 10.5 in S.I. (spark ignition) engine and 12 to 25 for C.I. (compression ignition) engine.

10.2.9 Mean Effective Pressure (Pm)

The mean effective pressure represents the theoretical constant pressure for which the workdone during volume change V_s is the same as net work.

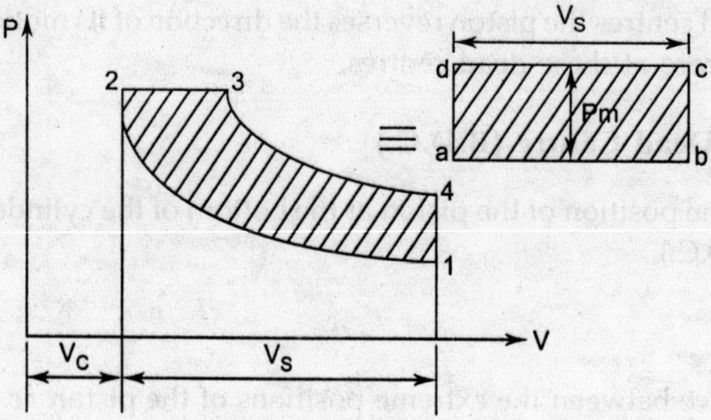

Fig. 10.1

$$W_{net} = P_m \times V_s$$

In Fig. 10.1, area $1-2-3-4-1$ = area $a-b-c-d-a$.

Thus $$P_m = \frac{W_{net}}{V_s} = \frac{\text{Net workdone per cycle}}{\text{swept volume}}.$$

So in otherwords the mean effective pressure is the theoretical constant pressure which, if it acted on the piston during power stroke, would produce the same net work as actually developed in one cycle.

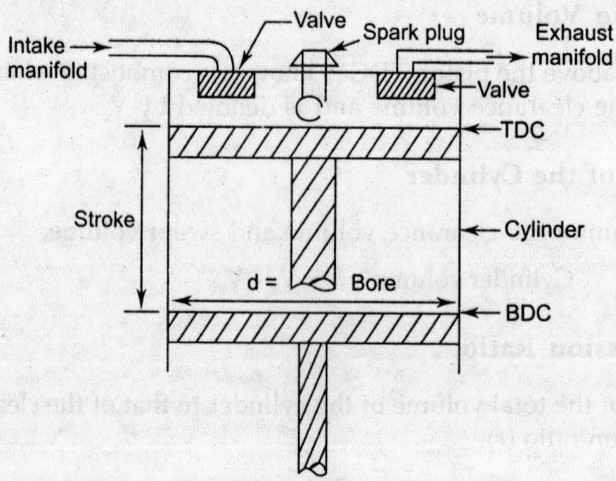

Fig. 10.2

10.3 OTTO CYCLE

This cycle is named after Dr. A.N.OTTO, a German scientist (in 1876).

The cycle that may be used to approximate the operation of S.I engine is the otto cycle. Fig. 10.3 shows the otto cycle on $P - v$ and $T - s$ diagram. Let 1 *kg* mass of ideal gas circulate in the cycle.

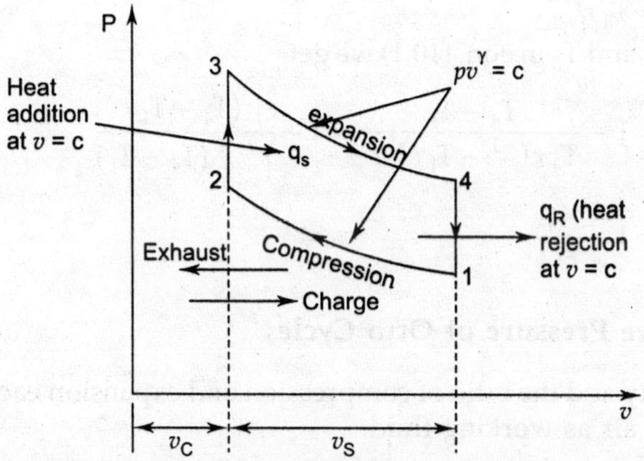

Fig. 10.3 (a)

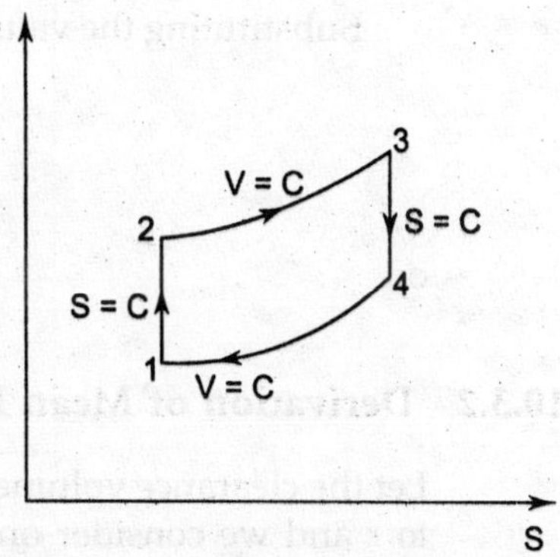

Fig. 10.3 (b)

Process	*Description*
1 – 2	Reversible adiabatic compression.
2 – 3	Constant volume heat addition.
3 – 4	Reversible adiabatic expansion.
4 – 1	Constant volume heat rejection.

10.3.1 Derivation of Air Standard Efficiency of Otto-cycle

$$q_s = \text{heat addition} = c_v\,(T_3 - T_2)$$

$$q_R = \text{heat rejection} = c_v\,(T_4 - T_1)$$

Efficiency
$$\eta_{otto} = 1 - \frac{q_R}{q_S} = 1 - \frac{c_v\,(T_4 - T_1)}{c_v\,(T_3 - T_2)} = 1 - \frac{T_4 - T_1}{T_3 - T_2} \quad ...(10.1)$$

$$r = \text{compression ratio} = \frac{v_1}{v_2}.$$

Expansion ratio
$$= \frac{v_4}{v_3} = \frac{v_1}{v_2} = r \quad [\text{as } v_4 = v_1, v_3 = v_2]$$

For process 1 – 2

$$P_1\,v_1^\gamma = P_2\,v_2^\gamma$$

or
$$P_1 v_1\,v_1^{\gamma-1} = P_2\,v_2\,v_2^{\gamma-1}$$

or
$$RT_1\,v_1^{\gamma-1} = RT_2\,v_2^{\gamma-1} \quad \left[r = \frac{v_1}{v_2}\right]$$

or
$$T_2 = T_1\,r^{\gamma-1} \quad ...(10.2)$$

For process 3 – 4

$$P_3 v_3^\gamma = P_4 v_4^\gamma$$

or
$$P_3 v_3\,v_3^{\gamma-1} = P_4\,v_4\,v_4^{\gamma-1}$$

or
$$RT_3\,v_3^{\gamma-1} = RT_4\,v_4^{\gamma-1} \quad \left[r = \frac{v_4}{v_3}\right]$$

or $$T_3 = T_4 r^{\gamma-1} \quad \text{...(10.3)}$$

Substituting the values of T_2 and T_3 in eqn. (10.1) we get

$$\eta_{otto} = 1 - \frac{T_4 - T_1}{T_4 r^{\gamma-1} - T_1 r^{\gamma-1}} = 1 - \frac{(T_4 - T_1)}{r^{\gamma-1}(T_4 - T_1)}$$

or $$\eta_{otto} = 1 - \frac{1}{r^{\gamma-1}} \quad \text{...(10.4)}$$

10.3.2 Derivation of Mean Effective Pressure of Otto Cycle

Let the clearance volume be unity and the ratio of compression and expansion each equal to r and we consider one *kg* of air as working fluid.

$$v_2 = v_3 = v_c = 1$$

Let $$\frac{P_3}{P_2} = \frac{P_4}{P_1} = \alpha$$

$\therefore$ $$P_1 v_1^\gamma = P_2 v_2^\gamma \text{ and } P_3 v_3^\gamma = P_4 v_4^\gamma$$

or $$\frac{P_2}{P_1} = \left(\frac{v_1}{v_2}\right)^\gamma \text{ and } \frac{P_3}{P_4} = \left(\frac{v_4}{v_3}\right)^\gamma$$

or $$\frac{P_2}{P_1} = r^\gamma \text{ and } \frac{P_3}{P_4} = r^\gamma.$$

Net workdone = area under the diagram

= (area under isentropic 3 – 4) – (area under isentropic 1 – 2)

$$= \int_3^4 P\,dv - \int_2^1 P\,dv \qquad [pv^\gamma = c]$$

$$= \int_3^4 \frac{c}{v^\gamma} dv + \int_1^2 \frac{c}{v^\gamma} dv$$

$$= \left.\frac{c v^{-\gamma+1}}{-\gamma+1}\right|_3^4 + \left.\frac{c v^{-\gamma+1}}{-\gamma+1}\right|_1^2$$

$$= \frac{c}{1-\gamma}\left[v_4^{1-\gamma} - v_3^{1-\gamma}\right] + \frac{c}{1-\gamma}\left[v_2^{1-\gamma} - v_1^{1-\gamma}\right]$$

$$= \frac{c v_4^{1-\gamma} - c v_3^{1-\gamma}}{1-\gamma} + \frac{c v_2^{1-\gamma} - c v_1^{1-\gamma}}{1-\gamma}$$

$$= \frac{P_4 v_4^\gamma v_4^{1-\gamma} - P_3 v_3^\gamma v_3^{1-\gamma}}{1-\gamma} + \frac{P_2 v_2^\gamma v_2^{1-\gamma} - P_1 v_1^\gamma v_1^{1-\gamma}}{1-\gamma}$$

$$= \frac{P_4 v_4 - P_3 v_3}{1-\gamma} + \frac{P_2 v_2 - P_1 v_1}{1-\gamma}$$

$$= \frac{P_3 v_3 - P_4 v_4}{\gamma - 1} - \frac{P_2 v_2 - P_1 v_1}{\gamma - 1}$$

$$= \frac{1}{\gamma-1}\left[P_4 r^\gamma - P_4 r - \left(P_1 r^\gamma - P_1 r\right)\right],$$

$$\left[\begin{array}{l}\textbf{Note:} \\ \text{Now } P_2 = P_1 r^\gamma \\ P_3 = P_4 r^\gamma \\ v_1 = v_4 = r \\ v_2 = v_3 = 1 \\ \dfrac{P_4}{P_1} = \alpha = \dfrac{P_3}{P_2}\end{array}\right]$$

$$= \frac{1}{\gamma-1}\left[P_4\left(r^\gamma - r\right) - P_1\left(r^\gamma - r\right)\right],$$

$$= \frac{1}{\gamma-1}\left[P_1 \alpha (r^\gamma - r) - P_1 (r^\gamma - r)\right],$$

$$= \frac{P_1}{\gamma-1}\left[\alpha r^{\gamma-1} - r\,(r^{\gamma-1} - 1)\right],$$

$$= \frac{P_1 r \left(r^{\gamma-1} - 1\right)(\alpha - 1)}{\gamma - 1},$$

$$w_{net} = \frac{P_1 r (\alpha - 1)\left(r^{\gamma-1} - 1\right)}{\gamma - 1},$$

$$\left[\begin{array}{l}\dfrac{v_c + v_s}{v_c} = r \\ v_s = (r-1)\,v_c \\ = (r-1) \textit{ as } v_c = 1\end{array}\right]$$

$$P_m = \frac{w_{net}}{v_s}$$

or

$$P_m = \frac{P_1 r (\alpha - 1)\left(r^{\gamma-1} - 1\right)}{(\gamma - 1)(r - 1)}.$$

Note : The Carnot efficiency based upon the maximum temperature of this cycle is equal to $\left(1 - \frac{T_1}{T_3}\right)$. Comparing this expression with the efficiency of the otto cycle *i.e.,* $\left(1 - \frac{T_4}{T_3}\right)$, we see that the efficiency of otto cycle is lower as $T_4 > T_1$. This is obvious also because in otto cycle all heat is not supplied at the highest temperature of the cycle *i.e.,* T_3 and not rejected at the lowest temperature of the cycle *i.e.,* T_4 as it is done in case of the Carnot cycle.

10.4 DIESEL CYCLE

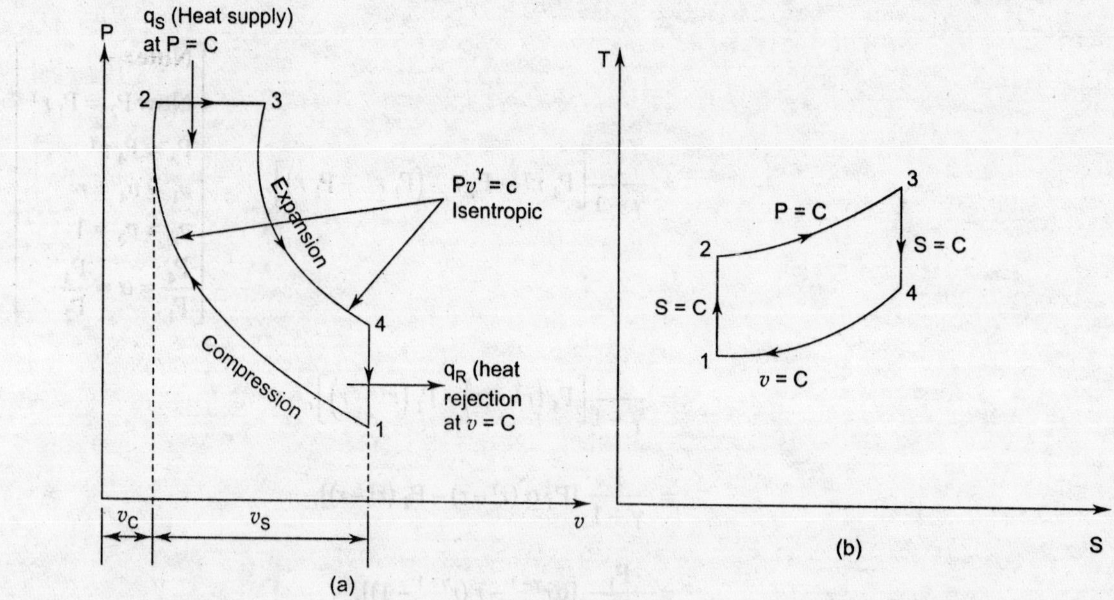

Fig. 10.4

Diesel cycle is named after Rudolph Diesel (1858 – 1913), a German engineer).

This cycle is used to approximate the operation of a compression ignition (C.I.) internal combustion engine.

Process	*Description*
1 – 2	Reversible adiabatic compression.
2 – 3	Constant pressure heat addition.
3 – 4	Reversible adiabatic expansion.
4 – 1	Constant volume heat rejection.

Let 1 *kg* of ideal gas circulate in the cycle.

10.4.1 Derivation of Air Standard Efficiency of Diesel Cycle

$$q_S = c_p (T_3 - T_2)$$
$$q_R = c_v (T_4 - T_1)$$

$$\eta_{Diesel} = 1 - \frac{q_R}{q_S} = 1 - \frac{c_v (T_4 - T_1)}{c_p (T_3 - T_2)} = 1 - \frac{T_4 - T_1}{\gamma (T_3 - T_2)}, \quad ...(10.5)$$

Process 1 – 2

$$P_1 v_1^\gamma = P_2 v_2^\gamma$$

or $$P_1 v_1 v_1^{\gamma-1} = P_2 v_2 v_2^{\gamma-1}$$

$$RT_1 v_1^{\gamma-1} = RT_2 v_2^{\gamma-1} \qquad \left[r = \frac{v_1}{v_2} = \textit{compression ratio} \right]$$

or $$T_2 = T_1 r^{\gamma-1} \quad ...(10.6)$$

Process 2 – 3

$$\frac{P_2 v_2}{T_2} = \frac{P_3 v_3}{T_3}, \qquad [P_2 = P_3]$$

or $$T_3 = T_2\left(\frac{v_3}{v_2}\right) = T_2 \beta, \qquad \left[\beta = \frac{v_3}{v_2} = \text{cut-off ratio}\right]$$

or $$T_3 = T_1 \beta r^{\gamma-1} \quad ...(10.7)$$

Process 3 – 4

$$P_3 v_3^{\gamma} = P_4 v_4^{\gamma}$$

$$\left[\begin{array}{l} v_4 = v_1 \\ \text{expansion ratio} \\ = \dfrac{v_4}{v_3} = \dfrac{v_4}{v_2}\cdot\dfrac{v_2}{v_3} \\ = \dfrac{\dfrac{v_1}{v_2}}{\dfrac{v_3}{v_2}} = \dfrac{r}{\beta} \end{array}\right.$$

or $$T_3 v_3^{\gamma-1} = T_4 v_4^{\gamma-1}$$

or $$T_4 = T_3\left(\frac{v_3}{v_4}\right)^{\gamma-1} = T_3\left(\frac{v_3}{v_1}\right)^{\gamma-1}$$

or $$T_4 = T_3\left(\frac{\beta}{r}\right)^{\gamma-1} = T_1 \beta r^{\gamma-1}\left(\frac{\beta}{r}\right)^{\gamma-1}$$

or $$T_4 = T_1 \beta r^{\gamma-1} . r^{1-\gamma} \beta^{\gamma-1} = T_1 \beta^{\gamma}$$

or $$T_4 = T_1 \beta^{\gamma} \quad ...(10.8)$$

Inserting the values of T_2, T_3 and T_4 in terms of T_1 from equations (10.6), (10.7) and (10.8) to equation (10.5) we get

$$\eta_{\text{Diesel}} = 1 - \frac{T_1 \beta^{\gamma} - T_1}{\gamma\left(T_1 \beta r^{\gamma-1} - T_1 r^{\gamma-1}\right)} = 1 - \frac{\beta^{\gamma} - 1}{\gamma r^{\gamma-1}(\beta - 1)}$$

10.4.2 Mean Effective Pressure (Pm)

It is easier to find out w_{net} from $q_S - q_R$ than by area under the curve since $w_{net} = w_{2-3} + w_{3-4} - w_{1-2}$ which leads to chances to commit mistakes.

$$\therefore \qquad Pm = \frac{w_{net}}{v_1 - v_2} = \frac{w_{net}}{(r-1)v_c} = \frac{q_S - q_R}{(r-1)v_c}.$$

10.5 OPERATING PRINCIPLE OF FOUR STROKE PETROL ENGINE (S.I. ENGINE)

The four stroke S.I. engine operates in a otto cycle which is air standard cycle. The four stages are explained below : (Ref Fig. 10.5 (*a*))

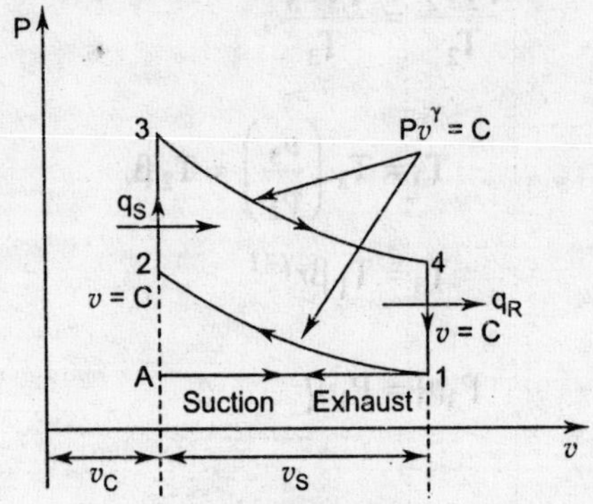

Fig. 10.5(a)

Stage 1 (Suction Stroke)

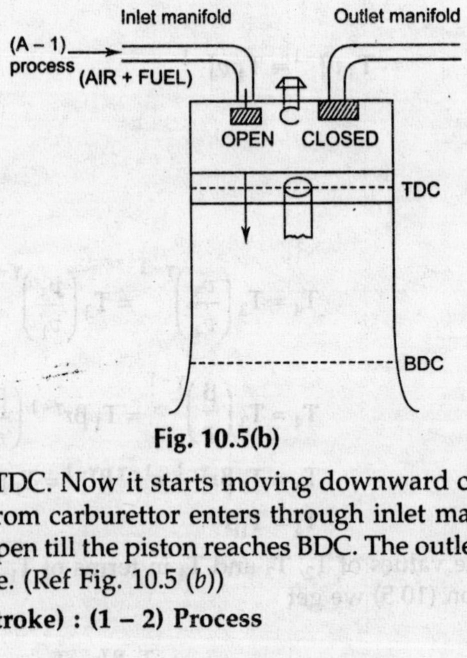

Fig. 10.5(b)

Suppose the piston is at TDC. Now it starts moving downward creating vacuum above it. The air fuel mixture from carburettor enters through inlet manifold to the cylinder. The inlet valve remains open till the piston reaches BDC. The outlet valve remains closed during this suction stroke. (Ref Fig. 10.5 (*b*))

Stage 2 (Compression Stroke) : (1 – 2) Process

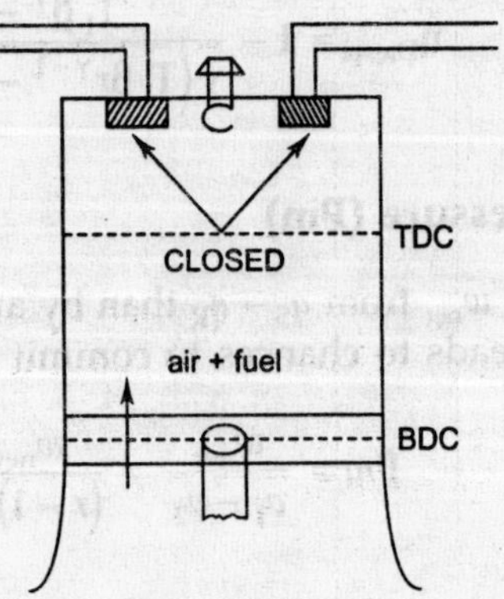

Fig. 10.5(c)

The piston starts moving upward from BDC. Inlet and outlet valves remain closed. The air fuel mixture is compressed with a compression ratio of 5 to 10.5 generally. The piston moves upward until it reaches TDC. The high pressure and high temperature (250°C to 500°C) are attained during this compression stroke. (Ref. Fig. 10.5 (*c*))

Stage 3 (Expansion Stroke or Working Stroke or Power Stroke) : (3–4 Process)

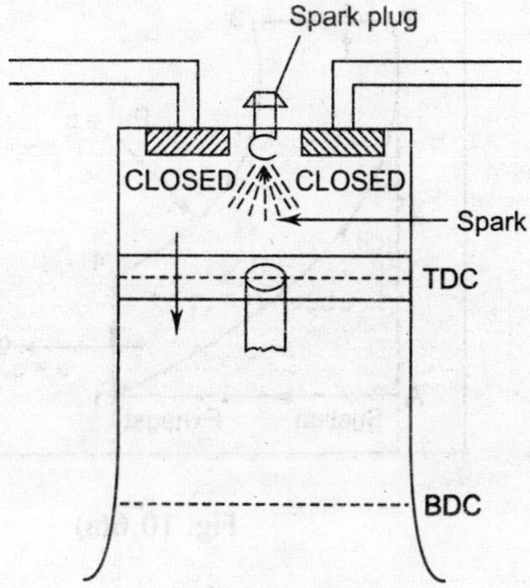

Fig. 10.5(d)

The charge is ignited by spark plug shortly before the piston reaches the TDC in the compression stroke (2–3 process). The product of combustion expands and give thrust to piston causing it to move downward. Both the valves are closed. We get power during this expansion stroke. The flywheel (not shown here) gets energy and provides other three strokes. At the end of expansion stroke heat is rejected at constant volume condition (4–1 process). (Ref. Fig. 10.5 (*d*))

Stage 4 (Exhaust Stroke) : (1–A) Process

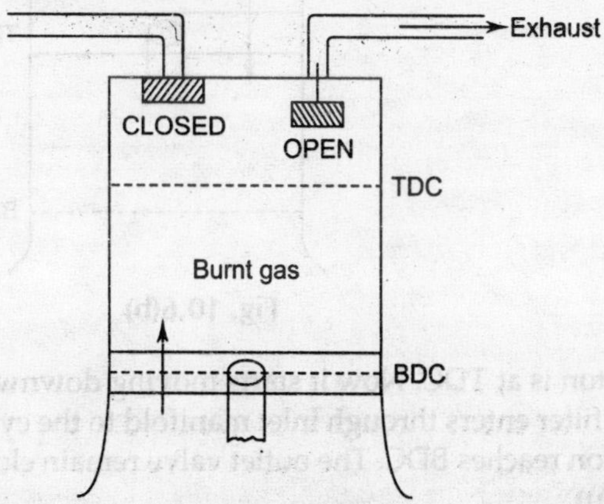

Fig. 10.5(e)

Now the exhaust valve opens and the burnt gases are removed from the cylinder. The piston moves upto TDC. The inlet valve remains closed. The process of removing gases is also called scavanging. (Ref. Fig. 10.5 (*e*))

10.6 OPERATING PRINCIPLE OF FOUR STROKE DIESEL ENGINE (C.I. ENGINE)

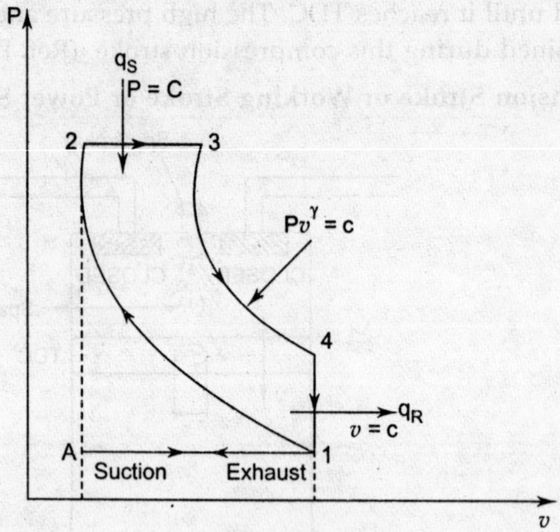

Fig. 10.6(a)

The four, stroke C.I. engine operates in a diesel cycle which is also an air standard cycle. The four stages are explained below. (Ref. Fig. 10.6 (*a*))

Stage 1 Suction Stroke (Process A–1)

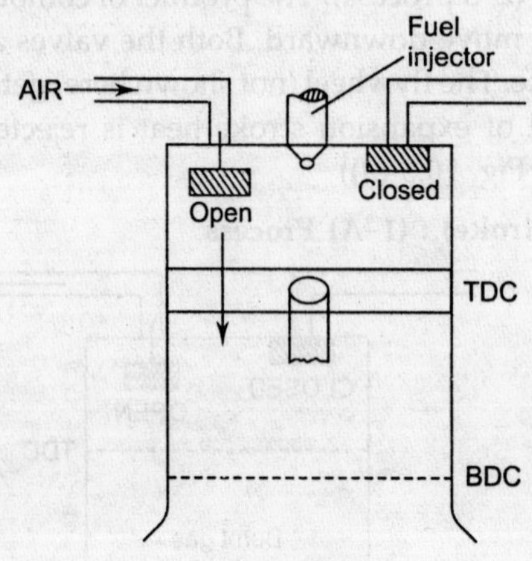

Fig. 10.6(b)

Suppose the piston is at TDC. Now it starts moving downward creating vacuum above it. The air from air filter enters through inlet manifold to the cylinder. The inlet valve remains open till the piston reaches BDC. The outlet valve remain closed during this suction stroke. (Ref. Fig. 10.6 (*b*))

Stage 2 (Compression Stroke) (Process 1–2)

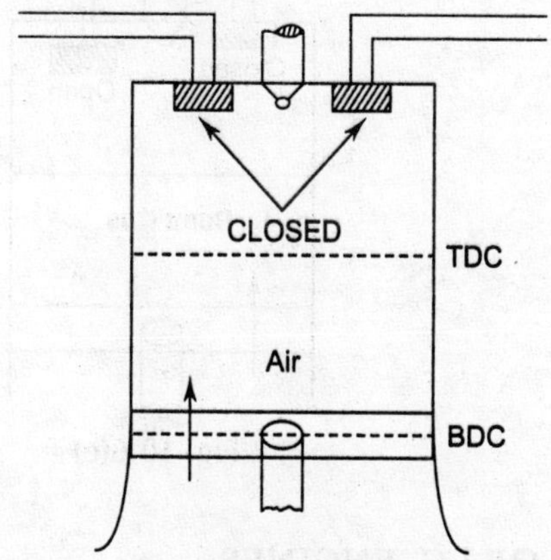

Fig. 10.6(c)

Both the valves remain closed; the piston moves from BDC to TDC. The air is compressed to a compression ratio of 12 to 25 generally. The air temperature reaches auto-ignite state. Now the fuel is injected from injector as soon as the piston reaches at TDC. (Ref. Fig. 10.6 (*c*))

Stage 3 Expansion Stroke (Process 3–4)

The fuel starts burning at constant pressure (Process 2–3). At point 3 fuel supply is cut off. The hot gases of the cylinder expands creating power stroke. Thus power is obtained by piston and flywheel during this expansion stroke. Both the valves remain closed during this process. The expansion is reversible adiabatic. At the end of this stroke the heat is rejected at constant volume condition (Process 4–1). (Ref. Fig. 10.6 (*d*))

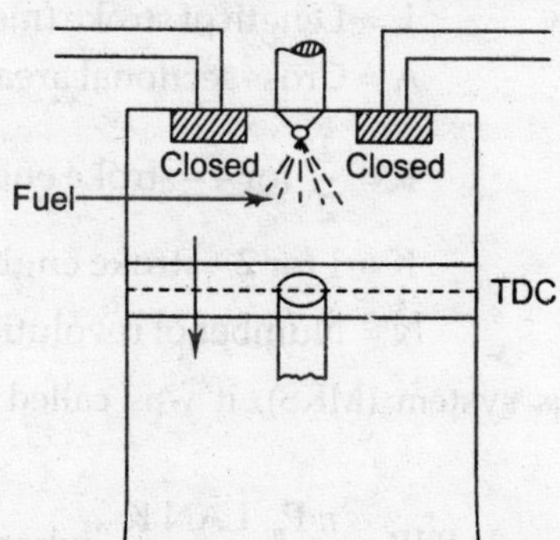

Fig. 10.6(d)

Stage 4 Exhaust Stroke (Process 1–A)

At this stroke the piston moves from BDC to TDC. The exhaust valve opens and inlet valve remain closed. The upward movement of the piston expels out the burnt gases (Process 1–A). (Ref. Fig. 10.6 (*e*))

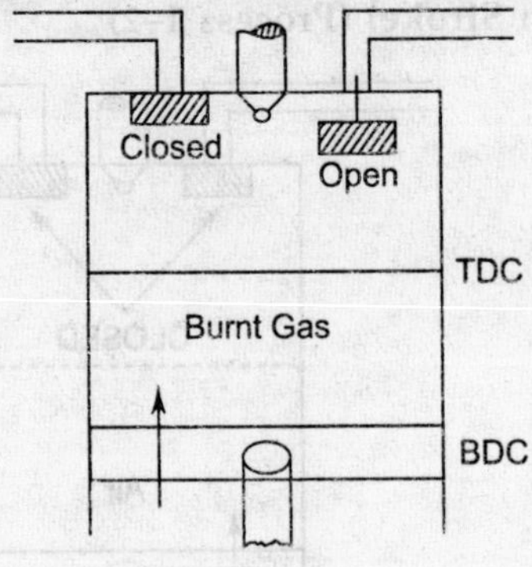

Fig. 10.6(e)

10.7 PERFORMANCE OF I.C. ENGINES

10.7.1 Indicated Power (IP)

The total power which is developed by combustion of fuel in the combustion chamber is called indicated power. It is given by

IP = Workdone by the cycle × no of cycles per second.

$$\therefore \quad IP = \frac{n P_m LANK}{60} \text{ (KW)}$$

where

n = Number of cylinders

P_m = Indicated mean effective pressure (KPa) = KN/m^2.

L = Length of stroke (metre)

A = Cross sectional area of piston (metre^2)

$K = \frac{1}{2}$ for 4 – stroke engine.

K = 1 for 2 – stroke engine.

N = Number of revolution of crank shaft/minute (*rpm*).

[Note that in previous system (MKS), it was called IHP (indicated horse power) 1HP = 746 Watt.

$$IHP = \frac{n P_m LANK}{4500}, \text{ where } P_m \text{ is in bars.}$$

10.7.2 Brake Power (BP)

The power developed by an engine at the output shaft is called brake power (BP). It is given by

$$BP = \frac{2\pi NT}{60 \times 1000} \text{ KW.}$$

where

N = Speed in rpm

T = Torque developed at shaft in Nm.

10.7.3 Frictional Power (FP)

If BP is subtracted from IP, it is frictional power (FP).

$$\therefore \qquad FP = IP - BP.$$

10.7.4 Mechanical Efficiency

The ratio of B.P to I.P is the mechanical efficiency.

$$\% \eta_{mech} = \frac{B.P}{I.P} \times 100$$

10.7.5 Volumetric Efficiency

The ratio of actual volume (reduced to NTP) of the charge, drawn in during suction stroke to swept volume of the piston is called volumetric efficiency.

10.7.6 Thermal Efficiency

The ratio of IP to energy supplied per second by the fuel. It is of two types.

10.7.6.1 Indicated Thermal Efficiency

$$\text{Indicated } \eta_{thermal} = \frac{IP}{\dot{m}_f \times C} = \frac{\text{workdone per cycle}}{\text{heat added per cycle}}.$$

where $\dot{m}_f$ = mass of fuel supplied in Kg per second

C = lower calorific value of the fuel

10.7.6.2 Brake Thermal Efficiency

$$\text{Brake } \eta_{thermal} = \frac{BP}{\dot{m}_f \times C}.$$

10.7.7 Specific Output

It is the brake power per unit piston displaced volume.

$$\text{specific output} = \frac{BP}{AL}.$$

where $A = \frac{\pi}{4} d^2$ in m^2

and d = bore diameter in m

L = stroke length in m.

SOLVED EXAMPLES

Examples Based on Otto Cycle

Example 10.1. *An engine working on ideal otto cycle has bore and stroke of the cylinder 17 cm and 30 cm respectively. The clearance volume is given 0.002025 m^3. Calculate the air standard efficiency of the engine. Assume $\gamma = 1.4$.*

Solution. Given

d = diameter of cylinder

= bore = 17 cm = 0.17 m

l = stroke length = 30 cm = 0.30 m

Clearance volume = $v_c = v_2 = 0.002025\ m^3$

$\gamma = 1.4, \quad \eta = ?$

$$v_s = \text{swept volume} = \frac{\pi}{4} d^2 l$$

$$= \frac{\pi}{4}(0.17)^2 \times 0.3\, m^3$$

$$= 6.81 \times 10^{-3}\ m^3$$

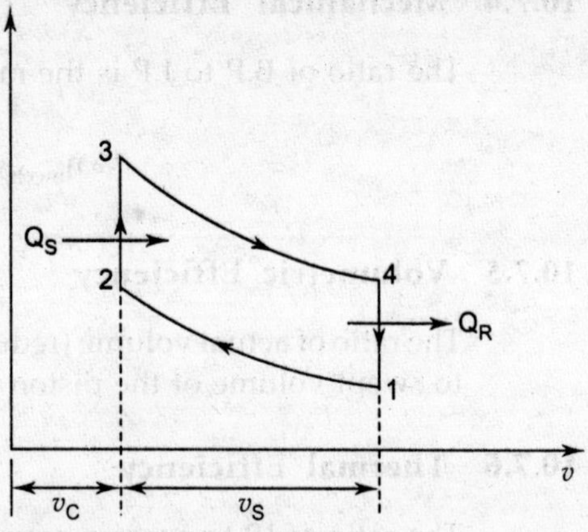

Fig. 10.7

$$v_1 = v_c + v_s = 0.002025 + 6.81 \times 10^{-3} = 8.835 \times 10^{-3}\ m^3$$

$$r = \frac{v_1}{v_2} = \frac{8.835 \times 10^{-3}}{0.002025} = 4.363$$

$$\eta_{cycle} = 1 - \frac{1}{r^{\gamma-1}} = 1 - \frac{1}{(4.363)^{1.4-1}} = 44.526\%. \quad \textbf{Ans.}$$

Example 10.2. *Calculate the efficiency of a four stroke cycle gas engine, assumed to be working on the constant volume cycle, of stroke 45 cm, piston diameter 30 cm, and a clearance of 11.4 litres (Take $\gamma = 1.4$)*

Solution. Given

$d = 0.3$ m

$l = 0.45$ m

$V_2 = V_c = 11.4 \times 10^{-3}\ m^3$

$$\left[\begin{aligned} 1\text{ litre} &= 1000\,\text{cm}^3 \\ &= 10^3 \times \frac{1}{10^6} m^3 \\ &= 10^{-3} m^3 \end{aligned}\right]$$

$$V_s = \frac{\pi}{4} d^2 l = \frac{\pi}{4}(0.3)^2 \times 0.45$$

$$= 0.0318\ m^3$$

$$\text{Total volume} = V_1 = V_s + V_c = 0.0318 + 11.4 \times 10^{-3} = 0.0432\ m^3$$

$$r = \frac{V_1}{V_2} = \frac{0.0432}{11.4 \times 10^{-3}} = 3.789$$

$$\eta_{cycle} = 1 - \frac{1}{r^{\gamma-1}} = 1 - \frac{1}{(3.789)^{1.4-1}} = 0.413 = 41.3\%. \quad \textbf{Ans.}$$

Example 10.3. *A four-cylinder petrol engine has a total swept volume of 3000 cm³ and clearance volume of each cylinder is 70 cm³. If the pressure and the temperature at the beginning of compression are 1.03 bar and 25°C and the maximum cycle temperature is 1500°C, calculate (i) the air standard efficiency (ii) all pressure and temperature of the cycles. (iii) the mean effective pressure. [Take $\gamma = 1.4$, $c_v = 0.718$ KJ/kg °K]*

R = 0.287

Solution. (*i*)

Swept volume per cylinder $= \dfrac{3000}{4} = 750 \text{ cm}^3$.

compression ratio $r = \dfrac{v_s + v_c}{v_c} = \dfrac{750+70}{70} = 11.714 = \dfrac{v_1}{v_2} = \dfrac{v_4}{v_3}$

$\eta_{\text{air standard}} = 1 - \dfrac{1}{r^{\gamma-1}} = 1 - \dfrac{1}{(11.714)^{0.4}} = 62.63\%$ **Ans (*i*).**

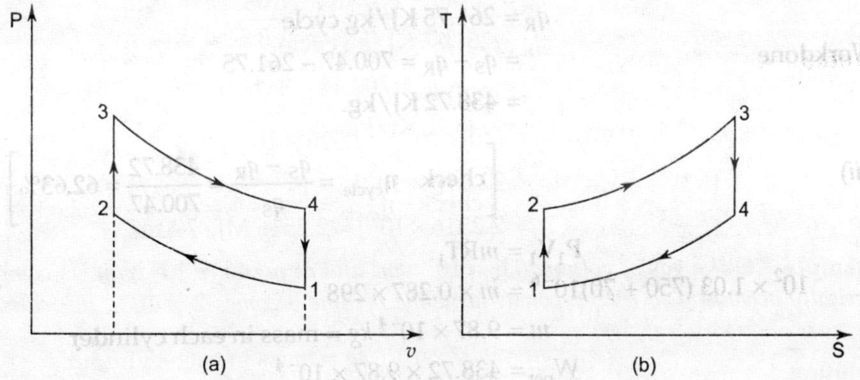

Fig. 10.8

(*ii*)

$$P_1 = 1.03 \text{ bar}$$
$$T_1 = 273 + 25 = 298°K$$
$$T_3 = 273 + 1500 = 1773°K$$
$$P_1 v_1^\gamma = P_2 v_2^\gamma$$
$$P_2 = P_1 \left(\frac{v_1}{v_2}\right)^\gamma = 1.03(11.714)^{1.4} = 32.286 \text{ bar} \quad \textbf{Ans.}$$

$$\frac{P_1 v_1}{T_1} = \frac{P_2 v_2}{T_2}$$

or $$T_2 = \frac{P_2}{P_1} \cdot \frac{v_2}{v_1} T_1 = \frac{32.286}{1.03} \times \frac{1}{11.714} \times 298 = 797.42°K \quad \textbf{Ans.}$$

$$\frac{P_2 v_2}{T_2} = \frac{P_3 v_3}{T_3} \text{ or } \frac{P_2}{T_2} = \frac{P_3}{T_3} \qquad [\text{as } v_2 = v_3]$$

or $$P_3 = \frac{T_3}{T_2} \cdot P_2 = \frac{1773}{797.42} \times 32.286 = 71.785 \text{ bar} \quad \textbf{Ans.}$$

$$P_3 v_3^\gamma = P_4 v_4^\gamma$$

or
$$P_4 = P_3\left(\frac{v_3}{v_4}\right)^\gamma = 71.785\left(\frac{1}{\frac{v_4}{v_3}}\right)^\gamma = \frac{71.785}{(11.714)^{1.4}}$$

$$= 2.29 \text{ bar} \quad \textbf{Ans.}$$

$$\frac{P_3 v_3}{T_3} = \frac{P_4 v_4}{T_4} \text{ or } T_4 = T_3\frac{P_4}{P_3}\left(\frac{v_4}{v_3}\right) = \frac{2.29}{71.785}\times 11.714\times 1773$$

$$= 662.55°K \quad \textbf{Ans.}$$

Heat supplied
$$= q_s = c_v(T_3 - T_2) = 0.718\,(1773 - 797.42)$$
$$= 700.47 \text{ KJ/kg cycle}$$

Heat rejected
$$q_R = c_v(T_4 - T_1) = 0.718\,(662.55 - 298)$$
$$q_R = 261.75 \text{ KJ/kg cycle}$$

Workdone
$$= q_S - q_R = 700.47 - 261.75$$
$$= 438.72 \text{ KJ/kg.}$$

(iii)
$$\left[\text{check } \eta_{cycle} = \frac{q_S - q_R}{q_S} = \frac{438.72}{700.47} = 62.63\%\right]$$

$$P_1V_1 = mRT_1$$
$$10^2 \times 1.03\,(750 + 70)10^{-6} = m \times 0.287 \times 298$$
$$m = 9.87 \times 10^{-4}\, kg = \text{mass in each cylinder}$$
$$W_{net} = 438.72 \times 9.87 \times 10^{-4}$$
$$= 0.433 \text{ KJ} \qquad [1 \text{ bar} = 10^5 \text{ N}/m^2 = 10^2 \text{ KN}/m^2]$$

$$Pm = \frac{W_{net}}{V_s} = \frac{0.433 \text{ KJ}}{750 \times 10^{-6}\, m^3}$$
$$= 577.33 \text{ KN}/m^2$$
$$= \frac{577.33}{10^2} \text{ bar} = 5.7733 \text{ bar.} \quad \textbf{Ans (iii).}$$

Example 10.4. *Show that for maximum work, the intermediate temperature in otto cycle is given by*

$$T_2 = T_4 = \sqrt{T_1 T_3} \quad where$$

T_1 and T_3 are initial maximum temperatures in compression and expansion process respectively.

Solution.

Workdone during the cycle $= \oint \delta w$

$$\oint dw = \oint \delta q$$

$$w = q_S - q_R$$

For Process 1–2 $\quad = C_v (T_3 - T_2) - C_v (T_4 - T_1)$

Now $\quad P_2 v_2^{\gamma} = P_1 v_1^{\gamma}$

or $\quad P_2 v_2 v_2^{\gamma-1} = P_1 v_1 v_1^{\gamma-1}$

or $\quad T_2 v_2^{\gamma-1} = T_1 v_1^{\gamma-1}$

or $$\frac{T_2}{T_1} = \left(\frac{v_1}{v_2}\right)^{\gamma-1} = r^{\gamma-1} \qquad ...(i)$$

Again for process (3 – 4) $\quad P_3 v_3^{\gamma} = P_4 v_4^{\gamma}$

or $\quad P_3 v_3 v_3^{\gamma-1} = P_4 v_4 v_4^{\gamma-1}$

or $\quad T_3 v_3^{\gamma-1} = T_4 v_4^{\gamma-1}$

or $$\frac{T_3}{T_4} = \left(\frac{v_4}{v_3}\right)^{\gamma-1} = r^{\gamma-1} \qquad ...(ii)$$

The following expression of w is a function of r when T_3 & T_1 are constant (fixed). The value of γ is constant

$$w = c_v (T_3 - T_1 r^{\gamma-1}) - c_v \left(\frac{T_3}{r^{\gamma-1}} - T_1\right)$$

$$= c_v T_3 - c_v T_1 r^{\gamma-1} - c_v T_3 r^{1-\gamma} + c_v T_1$$

$$\frac{\partial w}{\partial r} = 0 = -c_v T_1 (\gamma-1) r^{\gamma-2} - c_v T_3 (1-\gamma) r^{-\gamma}$$

or $$\frac{T_3}{T_1} = \frac{(1-\gamma) r^{\gamma-2}}{(1-\gamma) r^{-\gamma}} = r^{\gamma-2+\gamma} = r^{2(\gamma-1)} \qquad ...(iii)$$

or $$\sqrt{\frac{T_3}{T_1}} = r^{\gamma-1} = \frac{T_3}{T_4} \quad \Rightarrow \text{ (from } (ii) \text{ we get)}$$

or $$\frac{T_3}{T_1} = \frac{T_3^2}{T_4^2}$$

or $$T_4^2 = T_1 T_3 \qquad \left[\frac{T_2}{T_1} = \frac{T_3}{T_4} = r^{\gamma-1}\right]$$

or $$T_4 = \sqrt{T_1 T_3}.$$

And $$T_2 = \frac{T_3 T_1}{T_4} = \frac{T_3 T_1}{\sqrt{T_1 T_3}} = \sqrt{T_1 T_3}$$

$\therefore \quad T_2 = T_4 = \sqrt{T_1 T_3}$ **Proved.**

Example 10.5. *An engine working on the otto cycle is supplied with air at 0.1 MPa, 35°C. The compression ratio is 8. Heat supplied is 2100 KJ/kg. Calculate the maximum pressure and temperature of the cycle, the cycle efficiency and the mean effective pressure, (for air $c_p = 1.005$, $c_v = 0.718$, and $R = 0.287$ KJ/kg°k).*

Solution.

$$P_1 = 0.1 \text{ MPa} = 100 \text{ KN}/m^2$$

$$q_s = 2100 \text{ KJ}/kg = c_v (T_3 - T_2) \quad \text{...(i)}$$

$$T_1 = 273 + 35 = 308°K$$

$$r = 8 = \frac{v_1}{v_2}$$

$$P_3 = P_{max} = ?, T_3 = T_{max} = ?$$

$$\eta = ?, Pm = ?$$

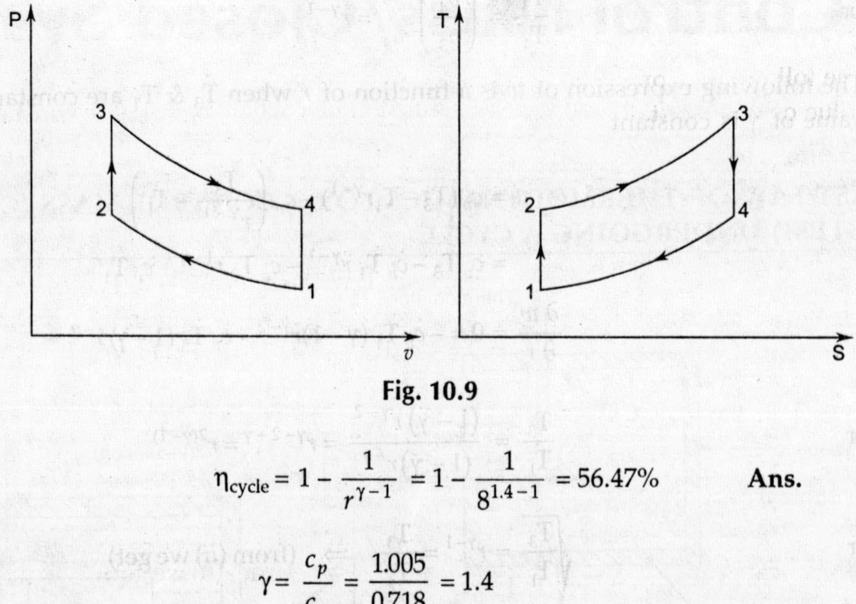

Fig. 10.9

$$\eta_{cycle} = 1 - \frac{1}{r^{\gamma-1}} = 1 - \frac{1}{8^{1.4-1}} = 56.47\% \quad \textbf{Ans.}$$

where $$\gamma = \frac{c_p}{c_v} = \frac{1.005}{0.718} = 1.4$$

$$\frac{P_1 v_1}{T_1} = \frac{P_2 v_2}{T_2} \text{ or } \frac{P_1}{T_1} = \frac{P_2}{T_2} \times \left(\frac{v_2}{v_1}\right) = \frac{P_2}{8 T_2}$$

or $$\frac{100}{308} = \frac{P_2}{8 T_2}$$

or $$\frac{P_2}{T_2} = 2.6 \quad \text{...(ii)}$$

(i) ⇒ $$2100 = 0.718 (T_3 - T_2)$$

or $$T_3 - T_2 = 2924.8°K \quad \text{...(iii)}$$

$$P_1 v_1^{\gamma} = P_2 v_2^{\gamma}$$

or $$P_2 = P_1\left(\frac{v_1}{v_2}\right)^{\gamma} = 100\,(8)^{1.4} = 1837.92\ KN/m^2$$

(ii) ⇒ $$T_2 = \frac{P_2}{2.6} = \frac{1837.92}{2.6} = 706.9°K \qquad ...(iv)$$

$$T_{max} = T_3 = 2924.8 + T_2 = 2924.8 + 706.9°$$
$$= 3631.7°K \textbf{ Ans.}$$

$$\frac{P_3 v_3}{T_3} = \frac{P_2 v_2}{T_2} \qquad [v_2 = v_3]$$

or $$P_3 = \left(\frac{P_2}{T_2}\right) T_3 = 2.6 \times 3631.7$$
$$= 9442.42\ KN/m^2$$
$$= 9.44242\ MPa \quad \textbf{Ans.}$$

$$w_{net} = q_s \times \eta = 2100 \times 0.5647$$
$$= 1185.87\ KJ/kg$$

$$P_m = \frac{w_{net}}{v_s} = \frac{w_{net}}{v_1 - v_2}$$

$$\frac{v_1}{v_2} = 8 \text{ and } P_1 v_1 = RT_1$$

or $$v_1 = \frac{RT_1}{P_1} = \frac{0.287 \times 308}{100}$$
$$= 0.884\ m^3/kg$$

$$v_2 = \frac{v_1}{8} = \frac{0.884}{8} = 0.11\ m^3/kg.$$

or $$P_m = \frac{w_{net}}{v_s} = \frac{w_{net}}{v_1 - v_2} = \frac{1185.87}{0.884 - 0.11} = 1532.13\ KPa$$
$$= 1.532\ MPa. \quad \textbf{Ans.}$$

Example 10.6. *In an otto cycle air at 1 bar, 17°C is compressed isentropically until the pressure is 15 bar. The heat is added at constant volume until the pressure rises to 40 bar. Calculate the air standard cycle efficiency and the mean effective pressure for the cycle. [Take c_v = 0.717 KJ/kg°K, R = 0.287 KJ/kg°K, γ = 1.4]*

Solution. Given

$$P_1 = 1 \text{ bar} = 10^5\ N/m^2$$
$$T_1 = 273 + 17 = 290°K$$

$$P_2 = 15 \text{ bar} = 15 \times 10^5 \text{ N}/m^2$$
$$P_3 = 40 \text{ bar} = 40 \times 10^5 \text{ N}/m^2$$
$$P_1 v_1^\gamma = P_2 v_2^\gamma$$

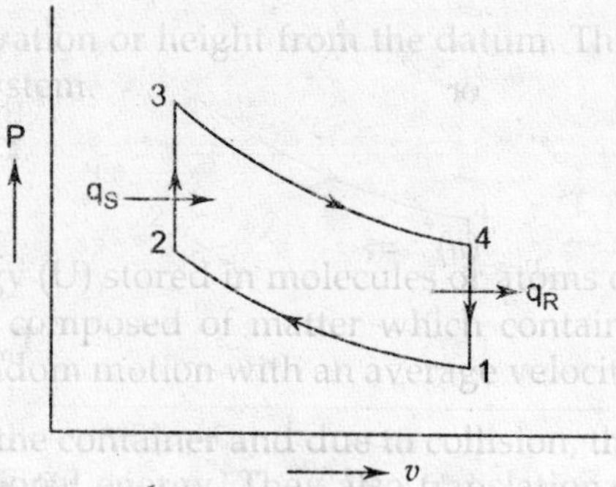

Fig. 10.10

or $$\frac{P_1}{P_2} = \left(\frac{v_2}{v_1}\right)^\gamma \Rightarrow \frac{v_1}{v_2} = \left(\frac{P_2}{P_1}\right)^{\frac{1}{\gamma}} = \left(\frac{15}{1}\right)^{\frac{1}{1.4}}$$
$$= 6.919 = r$$

$$\eta_{cycle} = 1 - \frac{1}{r^{\gamma-1}} = 1 - \frac{1}{(6.919)^{1.4-1}} = 53.87\% \quad \textbf{Ans.}$$

$$q_s = c_v (T_3 - T_2), \frac{P_3 v_3}{T_3} = \frac{P_2 v_2}{T_2} \qquad [v_2 = v_3]$$

or $$T_3 = \frac{P_3}{P_2} . T_2 = \frac{40}{15} \times T_2 = 2.67\, T_2$$

$$\frac{P_1 v_1}{T_1} = \frac{P_2 v_2}{T_2} \Rightarrow \frac{1 \times v_1}{290} = \frac{15 \times v_2}{T_2}$$

or $$T_2 = 15 \times 290 \times \frac{1}{\left(\frac{v_1}{v_2}\right)} = \frac{15 \times 290}{6.919} = 628.7°K$$

$\therefore$ $$T_3 = 2.67 \times T_2 = 2.67 \times 628.7 = 1678.63 \text{ °K}$$

$\therefore$ $$q_s = c_v (T_3 - T_2) = 0.717 (1678.63 - 628.7) = 752.8 \text{ KJ}/kg$$

$$w_{net} = \eta \times q_s$$
$$= 0.5387 \times 752.8 \text{ KJ}/kg = 405.53 \text{ KJ}/kg$$

$$P_m = \frac{w_{net}}{v_s}$$

$$\frac{v_1}{v_2} = 6.919, \frac{P_1 v_1}{T_1} = R \quad \text{where } m = 1 \text{ kg}$$

or $$v_1 = \frac{RT_1}{P_1} = \frac{0.287 \times 290}{100} = 0.8323\, m^3/\text{kg}.$$

$$v_2 = \frac{v_1}{6.919} = \frac{0.8323}{6.919} = 0.120\, m^3/kg.$$

$$P_m = \frac{405.53}{0.8323 - 0.12} = 569.3 \text{ KN}/m^2$$

$$= 5.693 \text{ bar } \textbf{Ans.}$$

Example 10.7. *An engine working on the otto cycle has a clearance of 17% of stroke volume and initial pressure of 0.95 bar and temperature 30°C. If the pressure at the end of constant volume heating is 28 bar. Find (i) Air standard efficiency (ii) the maximum temperature in the cycle (iii) Ideal mean effective pressure. Assume working fluid to be air.*

Solution. $\gamma = 1.4, c_p = 1.005\ KJ/kg°K, c_v = 0.714\ KJ/kg°K$

$v_c = 0.17\ v_s.$

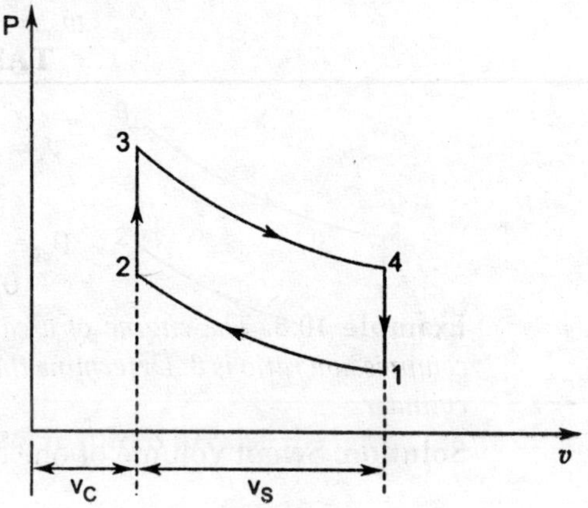

Fig. 10.11

$$r = \frac{v_1}{v_2} = \frac{v_c + v_s}{v_c} = 1 + \frac{v_s}{v_c}$$

or $$r = 1 + \frac{1}{0.17} = 6.88$$

$$\eta_{cycle} = 1 - \frac{1}{r^{\gamma-1}} = 1 - \frac{1}{(6.88)^{0.4}}$$

$= 53.76\%$ **Ans (*i*).**

$P_1 = 0.95\ bar = 95\ KN/m^2.$

$[1\ bar = 100\ KN/m^2]$

$T_1 = 273 + 30 = 303°K.$

$P_3 = 28\ bar = 2800\ KN/m^2.$

$$P_1 v_1^{\gamma} = P_2 v_2^{\gamma} \quad ...(1)$$

or $P_1 v_1\, v_1^{\gamma-1} = P_2 v_2\, v_2^{\gamma-1}$

or $T_1\, v_1^{\gamma-1} = T_2\, v_2^{\gamma-1}$

or $$\frac{T_1}{T_2} = \left(\frac{v_2}{v_1}\right)^{\gamma-1} = \left(\frac{1}{6.88}\right)^{0.4}$$

or $T_2 = T_1\,(6.88)^{0.4} = 303\,(6.88)^{0.4} = 655.35°K$

(1) $\Rightarrow$ $$P_2 = P_1 \left(\frac{v_1}{v_2}\right)^{\gamma} = 95\,(6.88)^{1.4} = 1413.66\ KN/m^2$$

$$\frac{P_2 v_2}{T_2} = \frac{P_3 v_3}{T_3} \quad [v_2 = v_3]$$

or $$T_3 = \frac{P_3}{T_2}.T_2 = \frac{2800}{1413.66} \times (655.35) = 1298°K.$$

Maximum temperature $= 1298°K, v_1 = \frac{RT_1}{P_1} = \frac{0.287 \times 303}{95} m^3 = 0.9153\,m^3/kg$ **Ans (*ii*).**

or $$v_2 = \frac{v_1}{6.88} = \frac{0.9153}{6.88} = 0.133\ m^3/Kg$$

$$q_s \text{ (heat supplied)} = c_v (T_3 - T_2)$$
$$= 0.717\,(1298 - 655.35)$$
$$= 460.78 \text{ KJ}/kg$$

Note
$$v_1 = 0.9153\ m^3/kg$$
$$v_2 = v_c = 0.133\ m^3/kg$$
$$w_{net} = \eta q_s = 0.5376 \times 460.78 \qquad [\text{Take } m = 1\,kg]$$
$$= 247.71 \text{ KJ}/kg$$
$$v_s = \frac{1}{0.17}\,v_c = \frac{0.133}{0.17} = 0.782\ m^3/kg$$
$$P_m = \frac{w_{net}}{v_s} = \frac{247.71}{0.782} = 316.76 \text{ KN}/m^2 = 3.1676 \text{ bar} \quad \textbf{Ans (ii).}$$

Example 10.8. *The engine of a car has four cylinders of 70 mm base and 75mm stroke. The compression ratio is 8. Determine the cubic capacity of the engine and the clearance volume of each cylinder.*

Solution. Swept volume of one cylinder is

$$V_s = \frac{\pi}{4} d^2 l = \frac{\pi}{4} \times (7)^2 \times 7.5 = 288.63\ cc$$

Cubic capacity of the engine

$$= \text{Total swept volume of all cylinders}$$
$$= 4 \times 288.63\ cc = 1154.52\ cc \quad \textbf{Ans.}$$

$$r = \frac{V_s + V_c}{V_c} = 8$$

or
$$\frac{288.63 + V_c}{V_c} = 8$$

or
$$V_c = \frac{288.63}{7} = 41.23\ cm^3 \text{ for each cylinder} \quad \textbf{Ans.}$$

Example 10.9. *An engine working on ideal otto cycle has temperature and pressure at the beginning of compression as 25°C and 1.5 bar respectively. Find the compression ratio if $\gamma = 1.4$ and thermal efficiency = 48%. Also find the temperature and pressure at the end of compression.*

Solution. Given $T_1 = 273 + 25 = 298$ °K, $P_1 = 1.5$ bar, $\gamma = 1.4$, $\eta = 0.48$

$$\eta = 1 - \frac{1}{r^{\gamma-1}}$$

or
$$0.48 = 1 - \frac{1}{r^{1.4-1}} \text{ or } r = 5.13 = \frac{v_1}{v_2}$$

We have $P_1 v_1^{\gamma} = P_2 v_2^{\gamma}$

or
$$P_1 v_1\, v_1^{\gamma-1} = P_2 v_2\, v_2^{\gamma-1}$$
$$T_1\, v_1^{\gamma-1} = T_2\, v_2^{\gamma-1}$$

or
$$\frac{T_1}{T_2} = \left(\frac{v_2}{v_1}\right)^{\gamma-1}, \text{ or } T_2 = T_1\, r^{\gamma-1}$$

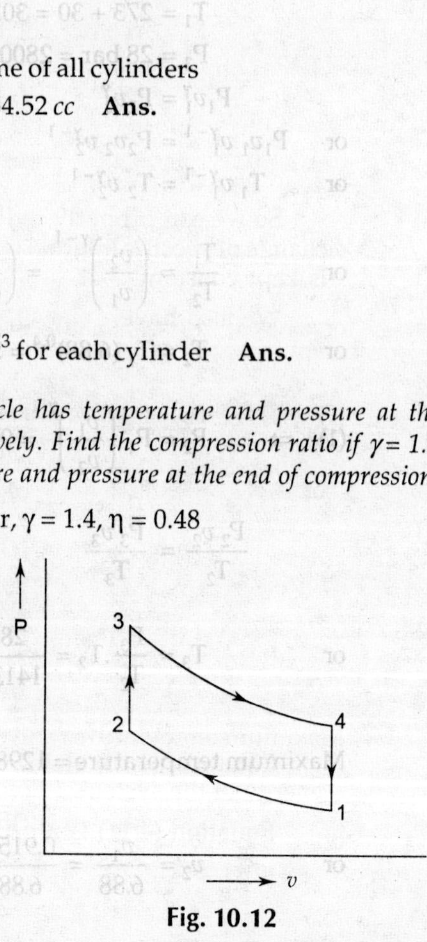

Fig. 10.12

$$= 298\,(5.13)^{0.4}$$

$$= 573°K = 300°C \quad \textbf{Ans.}$$

$$\frac{P_1 v_1}{T_1} = \frac{P_2 v_2}{T_2}$$

or $$\frac{P_1 T_2}{T_1}\left(\frac{v_1}{v_2}\right) = P_2$$

or $$1.5 \times \frac{573}{298} \times 5.13 = P_2$$

$$14.79 \text{ bar} = P_2 \quad \textbf{Ans.}$$

Check $$P_1 v_1^\gamma = P_2 v_2^\gamma$$

or $$P_2 = P_1\left(\frac{v_1}{v_2}\right)^\gamma$$

$$= 1.5\,(5.13)^{1.4}$$

$$= 14.79 \text{ bar} \quad \textbf{Ans.}$$

Example. 10.10. *An engine working on the otto cycle has a compression ratio of 8.5:1. The temperature and pressure at the beginning of compression are 93°C and 0.93 bar respectively. The maximum pressure in the cycle is 38 bar. Determine the pressure and temperature at the salient points of the cycle and the air standard efficiency.*

Solution. Given $$T_1 = 273 + 93 = 366°K,\ P_1 = 0.93 \text{ bar}$$

$$r = 8.5 = \frac{v_1}{v_2},\ P_3 = 38$$

$$P_2 v_2^\gamma = P_1 v_1^\gamma \text{ or } P_2 = P_1\left(\frac{v_1}{v_2}\right)^\gamma = 0.93\,(8.5)^{1.4}$$

$$= 18.606 \text{ bar } \textbf{Ans.}$$

$$P_2 v_2 v_2^{\gamma-1} = P_1 v_1 v_1^{\gamma-1}$$

or $$T_2 v_2^{\gamma-1} = T_1 v_1^{\gamma-1} \quad \text{or} \quad T_2 = T_1 r^{\gamma-1} = 366\,(8.5)^{1.4-1}$$

$$= 861.5°K \quad \textbf{Ans.}$$

At the beginning of expansion

Process 2 – 3 $$\frac{P_3 v_3}{T_3} = \frac{P_2 v_2}{T_2} \qquad [v_2 = v_3]$$

or $$T_3 = T_2 \cdot \frac{P_3}{P_2} = 861.5 \times \frac{38}{18.606} = 1759.5°K \quad \textbf{Ans.}$$

Process 3 – 4 $$P_3 v_3^\gamma = P_4 v_4^\gamma$$

or $$P_4 = P_3\left(\frac{v_3}{v_4}\right)^\gamma = P_3\left(\frac{v_2}{v_1}\right)^\gamma = \frac{P_3}{r^\gamma} = \frac{38}{8.5^{1.4}}$$

or $$P_4 = 1.89 \text{ bar.} \quad \textbf{Ans.}$$

Process 4 – 1 $$\frac{P_4 v_4}{T_4} = \frac{P_1 v_1}{T_1} \qquad [v_1 = v_4]$$

or $$T_4 = T_1\left(\frac{P_4}{P_1}\right) = 366\left(\frac{1.89}{0.93}\right) = 743.81°K \quad \textbf{Ans.}$$

$$\eta_{\text{air stand}} = 1 - \frac{1}{r^{\gamma-1}} = 1 - \frac{1}{(8.5)^{1.4-1}} = 57.51\%. \quad \textbf{Ans.}$$

Example 10.11. *The compression ratio of an otto cycle is 8. At the beginning of compression the pressure and temperature are 1 bar and 300°K respectively. The heat transfer to the air per cycle is 1900 KJ/kg of air. Calculate (a) the pressure and temperature at the end of each process of the cycle (b) thermal efficiency and (c) the mean effective pressure. Take $c_p = 1.005$ KJ/kg°K, $c_v = 0.718$ KJ/kg°K, R = 0.287 KJ/kg°K.*

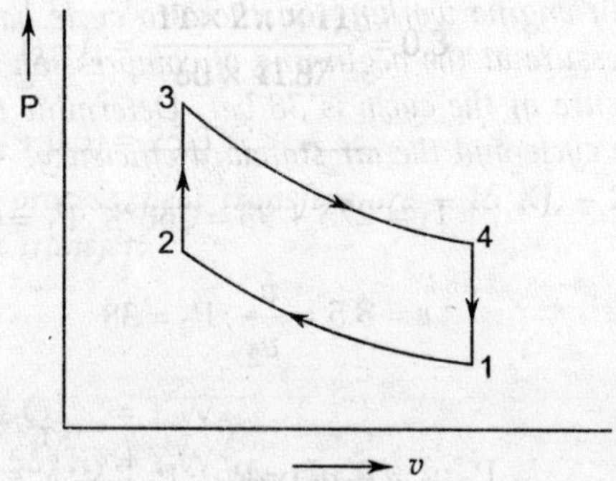

Fig. 10.13

Solution. (*a*) $$P_1 = 1 \text{ bar}, T_1 = 300°K, \gamma = \frac{c_p}{c_v} = \frac{1.005}{0.718} = 1.4$$

$$P_1 v_1 = RT_1 \text{ or } v_1 = \frac{RT_1}{P_1} = \frac{(0.287\times10^3)\times300}{1\times10^5} = 0.861\, m^3/kg$$

Process 1 – 2 $$\frac{T_2}{T_1} = r^{\gamma-1} \Rightarrow T_2 = T_1 r^{\gamma-1} = 300\,(8)^{0.4} = 689.22°K \quad \textbf{Ans.}$$

$$P_1 v_1^\gamma = P_2 v_2^\gamma \Rightarrow P_2 = P_1 (r)^\gamma = 1\times8^{1.4} = 18.38 \text{ bar} \quad \textbf{Ans.}$$

$$\frac{v_1}{v_2} = r \text{ or } v_2 = \frac{v_1}{r} = \frac{0.861}{8} = 0.1076\, m^3/kg.$$

Heat added per *kg* during constant volume. Process 2 – 3 is given by $(q_s) = c_v\,(T_3 - T_2)$

$$1900 = 0.718\,(T_3 - 689.22)$$

$$\Rightarrow \quad T_3 = 3335.46°K$$

Process 2 – 3 $$v_3 = v_2, P_3 = P_2 \times \frac{T_3}{T_2} = 18.38 \times \frac{3335.46}{689.22} = 88.95 \text{ bar} \quad \textbf{Ans.}$$

Process 3 – 4 $$T_4 = \frac{T_3}{r^{\gamma-1}} = \frac{3335.46}{8^{0.4}} = 1451.84°K$$

Process 3 – 4 $P_4 = \frac{P_3}{r^\gamma} = \frac{88.95}{8^{1.4}} = 4.839$ bar. **Ans.**

(*b*) $\eta = 1 - \frac{1}{r^{\gamma-1}} = 1 - \frac{1}{8^{0.4}} = 56.47\%$ **Ans.**

Heat rejected per *kg* of air during constant volume process (4 – 1) is given by

$$q_R = c_v (T_4 - T_1)$$
$$= 0.718 (1451.84 - 300)$$
$$= 827.02 \text{ KJ}/kg.$$

Net workdone $= q_s - q_R = (1900 - 827.02)$
$$= 1072.98 \text{ KJ}/kg$$

(*c*) Mean effective pressure

$$P_m = \frac{\text{Net workdone}}{\text{Stroke volume}} = \frac{1072.98}{v_1 - v_2}$$

$$= \frac{1072.98}{0.861 - 0.1076} \text{ KN}/m^2$$

$$= 1424.18 \text{ KN}/m^2 \qquad [1\text{bar} = 10^5 \text{ N}/m^2]$$

$$= 14.2418 \text{ bar. } \textbf{Ans.}$$

Example 10.12. *An engine working on otto cycle has a volume of 0.5 m^3, pressure 1 bar and temperature 27°C at the beginning of compression stroke. The pressure at the end of compression is 10 bar. The heat added during constant volume process is 200 KJ. Calculate (a) percentage clearance (b) efficiency and (c) mean effective pressure.*

Solution. $P_1 = 1$ bar, $P_2 = 10$ bar, $V_1 = 0.5\ m^3$, $T_1 = 273 + 27 = 300°K$

$1 - 2,\ P_1 v_1^\gamma = P_2 v_2^\gamma$

or $$\left(\frac{v_1}{v_2}\right)^\gamma = \frac{P_2}{P_1}$$

or $$r^\gamma = \frac{P_2}{P_1} \text{ or } r = \left(\frac{P_2}{P_1}\right)^{\frac{1}{\gamma}} = \left(\frac{10}{1}\right)^{\frac{1}{1.4}}$$

or $$r = 5.18$$

$$T_2 = T_1 r^{\gamma - 1} = 300 (5.18)^{0.4} = 579.23°K$$

$$\frac{P_1 V_1}{T_1} = \frac{P_2 V_2}{T_2} \text{ or } V_2 = \frac{P_1 V_1}{T_1} \times \frac{T_2}{P_2}$$

or $$V_2 = \frac{1 \times 0.5 \times 579.23}{300 \times 10} = 0.0965\ m^3$$

Heat supplied $= Q_S = m\, c_v (T_3 - T_2)$ ...(1)

where $$m = \frac{P_1 V_1}{RT_1} = \frac{1 \times 10^5 \times 0.5}{0.287 \times 10^3 \times 300}$$

$$= 0.58\ kg.$$

(1) ⇒ $Q_S = 200$ KJ

or $200 = 0.58 \times 0.718 \times (T_3 - 579.23)$

or $T_3 = 1059.5°K$

$$\frac{T_4}{T_3} = \left(\frac{v_3}{v_4}\right)^{\gamma-1} = \left(\frac{v_2}{v_4}\right)^{\gamma-1} = \left(\frac{1}{r}\right)^{\gamma-1}$$

or $$T_4 = T_3\left(\frac{1}{r}\right)^{\gamma-1} = 1059.5\left(\frac{1}{5.18}\right)^{1.4-1}$$

$$= 548.74°K.$$

(*a*) Percentage clearance $= \dfrac{V_2}{V_1 - V_2} \times 100$

$$= \frac{0.0965 \times 100}{0.5 - 0.0965} = 23.915\%. \quad \textbf{Ans.}$$

(*b*) $$\eta = 1 - \frac{1}{r^{\gamma-1}} = 1 - \frac{1}{(5.18)^{0.4}} = 48.21\% \quad \textbf{Ans.}$$

Heat rejected $Q_R = m\,c_v\,(T_4 - T_1)$

$$= 0.58 \times 0.718\,(548.74 - 300)$$

$$= 103.58 \text{ KJ}$$

Net workdone $= Q_S - Q_R = 200 - 103.58 = 96.42$ KJ.

(*c*) Mean effective pressure $= \dfrac{W_{net}}{V_s}$

$$= \frac{96.42 \times 10^3}{(V_1 - V_2)}\ \frac{Nm}{m^3}$$

$$= \frac{96.42 \times 10^3}{0.5 - 0.0965}\ N/m^2$$

$$= 238959.1\ N/m^2 \qquad [1 \text{ bar} = 10^5\ N/m^2]$$

$$= 2.389591 \text{ bar} \quad \textbf{Ans.}$$

Example 10.13. *The upper and lower temperature limits for an otto cycle are 1400°K and 300°K. Find the maximum theoretical power developed by the engine working on this cycle when the rate of flow of air is 0.33 kg/min.*

Solution. The condition for maximum work is given by

$$T_2 = T_4 = \sqrt{T_1\,T_3}$$

$$= \sqrt{300 \times 1400} = 648°K$$

Workdone per *kg* $= c_v\,[T_3 - T_2] - c_v\,[T_4 - T_1]$

$$= 0.718\,(1400 - 648) - 0.718\,(648 - 300)$$

$$w_{net} = 290.072 \text{ KJ}/kg.$$

∴ Maximum power developed

$$= 290.072 \frac{KJ}{kg} \times \frac{0.33\,kg}{60\,sec}$$

$$= 1.594\,KW. \quad \textbf{Ans.}$$

Example 10.14. *The bore and stroke of an engine working on the otto cycle are 17 cm and 30 cm respectively. The clearance volume is 0.001025 m^3. Calculate the air standard efficiency.*

Solution.

$$V_1 - V_2 = \text{swept volume} = V_s = \frac{\pi}{4} d^2 l = \frac{\pi}{4} \times 17^2 \times 30 = 6809\,cc$$

$$V_2 = \text{clearance volume} = V_c = 0.001025 \times 10^6\,cc = 1025\,cc$$

compression ratio $= r = \dfrac{V_1}{V_2} = \dfrac{V_c + V_s}{V_c} = \dfrac{1025 + 6809}{1025}$

$$= 7.64.$$

∴ Air standard efficiency $\eta = 1 - \dfrac{1}{r^{\gamma-1}} = 1 - \dfrac{1}{7.64^{1.4-1}}$

$$\eta = 55.7\%. \quad \textbf{Ans.}$$

Example 10.15. *In an ideal constant volume cycle the pressure and temperature at the beginning of compression are 97 KN/m^2 and 40°C, respectively. The volume ratio of compression is 7:1. The heat supplied during the cycle is 1200 KJ/kg of working fluid. Determine (a) the maximum temperature attained in the cycle (b) the thermal efficiency of the cycle and (c) the workdone during the cycle/kg of working fluid.*

Assume $\gamma = 1.4,\ c_v = 0.718$ KJ/kg°K

Solution. (*a*) $T_2 = T_1 r^{\gamma-1} = (273 + 40) 7^{1.4-1} = 681.7$ °K

$$q_S = c_v (T_3 - T_2)$$

$$1200 = 0.718 (T_3 - 681.7), \Rightarrow T_3 = 2353\,K = T_{max}. \quad \textbf{Ans.}$$

(*b*) $\eta = 1 - \dfrac{1}{r^{\gamma-1}} = 1 - \dfrac{1}{7^{1.4-1}} = 54.1\%$ **Ans.**

(*c*) Workdone/*kg*/cycle = heat supplied/*kg*/cycle × η

$$= 1200 \times 0.541 = 649\,KJ.$$

Alternatively:

workdone $= q_S - q_R$

$$= c_v (T_3 - T_2) - c_v (T_4 - T_1)$$

Now $T_3 = T_4 r^{\gamma-1}$

or $T_4 = \dfrac{T_3}{r^{\gamma-1}} = \dfrac{2353}{7^{1.4-1}} = 1080\,K.$

∴ workdone $= 1200 - 0.718 (1080 - 313)$

$$= 649\,KJ. \quad \textbf{Ans.}$$

Example 10.16. *In an ideal otto cycle the compression ratio is 8. The initial pressure and temperature of the air are 1 bar and 100°C. The maximum pressure in the cycle is 50 bar. For*

1 kg of air flow, calculate the values of the pressure, volume, and temperature at the four salient points of the cycle. What is the ratio of heat supplied to the heat rejected?

for air $R = 0.287$ *KJ/kg°K,* $\gamma = 1.4$.

Solution. $P_1 = 1$ bar

$T_1 = 273 + 100 = 373°K$ [1 bar = 100 KN/m^2]

$P_1V_1 = mRT_1$ $m = 1$kg

or $$V_1 = \frac{1 \times 0.287 \times 373}{1 \times 100} = 1.0705\ m^3 = v_1 \text{ as } m = 1\text{Kg}$$

Process 1 – 2 $P_1 v_1^{\gamma} = P_2 v_2^{\gamma}$

or $$P_2 = P_1 \left(\frac{v_1}{v_2}\right)^{\gamma} = 1 (8)^{1.4} = 18.38 \text{ bar} \quad \textbf{Ans.}$$

$$v_2 = \frac{v_1}{8} = \frac{1.0705}{8} = 0.1338\ m^3.$$

$$\frac{P_1 v_1}{T_1} = \frac{P_2 v_2}{T_2}$$

or $$T_2 = \frac{P_2}{P_1} \cdot \frac{v_2}{v_1} \cdot T_1 = \frac{18.38 \times 0.1338 \times 373}{1 \times 1.0705} = 856.8°K \quad \textbf{Ans.}$$

Process 2 – 3 $v_3 = v_2 = 0.1338\ m^3$, $P_3 = 50$ bar.

$$\frac{P_3 v_3}{T_3} = \frac{P_2 v_2}{T_2}, \text{ or } T_3 = \frac{P_3}{P_2} T_2$$

or $$T_2 = \frac{P_2}{P_3} \cdot T_3 = \frac{18.38}{50} \times T_3$$

or $$T_3 = \frac{P_3}{P_2} T_2 = \frac{50}{18.38} \times 856.8 = 2336.4°K \quad \textbf{Ans.}$$

Process 3 – 4 $P_3 v_3^{\gamma} = P_4 v_4^{\gamma}$

or $$P_4 = P_3 \left(\frac{v_3}{v_4}\right)^{\gamma} = 50 \left(\frac{1}{8}\right)^{1.4} = 2.72 \text{ bar}$$

$$\frac{P_4 v_4}{T_4} = \frac{P_1 v_1}{T_1}, \qquad [v_4 = v_1]$$

or $$T_4 = \frac{P_4}{P_1} \cdot T_1 = \frac{2.72}{1} \times 373 = 1014.6°K \quad \textbf{Ans.}$$

$$c_v = \frac{R}{\gamma - 1} = \frac{0.287}{1.4 - 1} = 0.718 \text{ KJ}/kg°K$$

$$q_S = c_v(T_3 - T_2) = 0.718\,(2336.4 - 856.8)$$
$$= 1062.4 \text{ KJ}/kg$$
$$q_R = c_v(T_4 - T_1) = 0.718\,(1014.6 - 373) = 460.7 \text{ KJ}/kg.$$

or, $$\frac{q_s}{q_R} = \frac{\text{Heat supplied}}{\text{Heat rejected}} = \frac{1062.4}{460.7} = 2.306 \quad \textbf{Ans.}$$

Points	*Pressure bar*	*Volume m^3*	*Temp °C*
1	1.0	1.0705	100
2	18.4	0.1338	583.8
3	50	0.1338	2063.4
4	2.72	1.0705	741.6

Example 10.17. *A four stroke engine having swept volume of 0.13 m^3, operates on otto cycle. The compression ratio is 6 and the conditions at the beginning of compression are 1 kg f/cm^2 and 60°C. The heat supplied is 36 Kcal/Kg/cycle. Calculate the work done, efficiency. Take c_p = 0.237 Kcal/kg K, c_v = 0.169 Kcal/kg K. γ = 1.4, (air)*

Solution. $$v_s = 0.13\,m^3,\ T_1 = 60°C = 60 + 273 = 333°K$$
$$r = 6,\ P_1 = 1\,kg\,f/\text{cm}^2,\ q_S = 36 \text{ kcal/cycle}$$
$$r = \frac{v_1}{v_2} = \frac{v_c + v_s}{v_c}$$

or $$6 = \frac{v_c + 0.13}{v_c}, \text{ or } v_c = 0.026\,m^3$$
$$v_1 = v_c + v_s = 0.156\,m^3$$
$$v_2 = v_c = 0.026\,m^3$$

Process 1 – 2 $$T_2 = T_1\,r^{\gamma-1} = 333\,(6)^{0.4} = 681.88°K$$
$$q_s = c_v(T_3 - T_2)$$
or $$36 = 0.169\,(T_3 - 681.88)$$
or $$T_3 = 894.89°K$$

Process 3 – 4 $$T_4 = \frac{T_3}{r^{\gamma-1}} = \frac{894.89}{6^{0.4}} = 437°K$$
$$q_R = c_v(T_4 - T_1) = 0.169\,(437 - 333) = 17.5 \text{ kcal}/kg$$

Workdone $$= q_S - q_R = 36 - 17.5 = 18.5 \text{ kcal}/kg$$
$$= 18.5 \times 4.184 \text{ KJ}/kg$$
$$= 77.4 \text{ KJ}/kg \quad \textbf{Ans.}$$ [1 Kcal = 4.184 KJ]

$$\eta = 1 - \frac{q_R}{q_s},\ 1 - \frac{17.5}{36} = 51.38\% \quad \textbf{Ans.}$$

Examples Based on Diesel Cycle

Example 10.18. *1 kg of air is taken through a diesel cycle. Initially the air is at 15°C and 1 bar. The compression ratio is 15 and the heat added is 1850 KJ. Calculate the ideal cycle efficiency and mean effective pressure.*

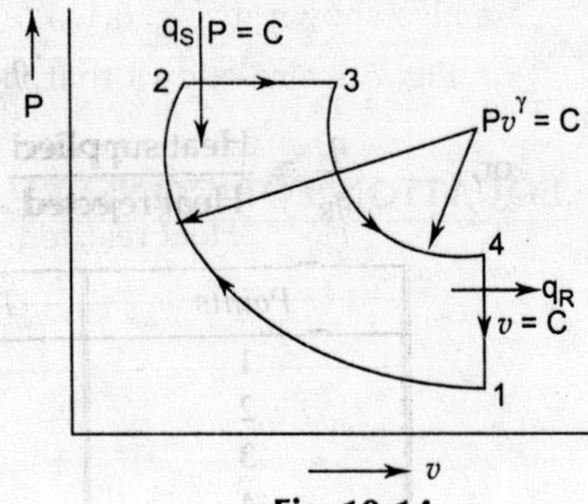

Fig. 10.14

Solution. $v_1 = \dfrac{RT_1}{P_1} = \dfrac{0.287 \times (15 + 273)}{100}$

$$= 0.81\ m^3/kg \left[1 \text{ bar} = 100\ \frac{KN}{m^2}\right]$$

$$T_2 = T_1\, r^{\gamma - 1} = 288\ (15)^{1.4-1} = 851\ °K \qquad \text{....(1)}$$

$$v_2 = \frac{v_1}{r} = \frac{0.81}{15} = 0.054\ m^3/Kg$$

$$2-3, \text{ Heat Added} = q_s = c_p\,(T_3 - T_2) \qquad \text{....(2)}$$

or $$1850 = 1 \times 1.005\ (T_3 - 851)$$

or $$T_3 = 2692\ K.$$

$$\left[\begin{aligned} &\text{Use } T_2 = T_1 r^{\gamma-1} \\ &T_4 = T_1\, \beta^{\gamma} \\ &T_3 = T_1\, \beta r^{\gamma - 1} \end{aligned}\right]$$

Process 2 – 3 $$\frac{P_2\,v_2}{T_2} = \frac{P_3\,v_3}{T_3}, \quad [P_2 = P_3]$$

or $$\frac{v_2}{T_2} = \frac{v_3}{T_3}$$

or $$\frac{T_3}{T_2} = \frac{v_3}{v_2} = \beta$$

or $$\beta = \frac{T_3}{T_2} = \frac{2692}{851} = 3.1633$$

or $$\frac{v_4}{v_3} = \frac{v_1}{v_2} \times \frac{v_2}{v_3} = \frac{r}{\beta} = \frac{15}{3.1633} = 4.742$$

$$T_4 = T_3 \left(\frac{v_3}{v_4}\right)^{\gamma - 1} = 2692 \left(\frac{1}{4.742}\right)^{0.4} = 1444°K$$

Head rejected $$q_R = c_v\,(T_4 - T_1) = 1 \times 0.718\ (1444 - 288)$$

$$= 830\ KJ/kg$$

$$\text{workdone} = \text{Heat added} - \text{Heat rejected}$$

$$= 1850 - 830 = 1020\ KJ/kg$$

$$\eta = \frac{\text{workdone}}{q_s} = \frac{1020}{1850} = 55.1\% \quad \textbf{Ans.}$$

Mean effective pressure $= P_m$

$$P_m = \frac{\text{Net workdone}}{v_1 - v_2} = \frac{1020}{0.81 - 0.054}\ KN/m^2$$

$$= \frac{1020}{(0.81 - 0.054)\ 100} \text{ bar}$$

$= 13.4$ bar. **Ans.**

Example 10.19. *In a diesel cycle the maximum pressure reached is 36 bar. The compression ratio is 13. The temperature at the beginning of compression is 83°C and the temperature at the beginning of expansion is 1800°C. Determine (a) the temperature at the end of compression (b) the temperature at the end of expansion (c) the heat supplied and heat rejected per kg of working fluid, and (d) the thermal efficiency of the cycle. Take c_p = 0.986 KJ/kg°K*

c_v = 0.702 KJ/kg°K.

Solution. Given $T_1 = 83°C = 83 + 273 = 356°K, r = 13,$

$T_3 = 1800°C = 1800 + 273 = 2073°K,$

$$\gamma = \frac{c_p}{c_v} = \frac{0.986}{0.702} = 1.4$$

(*a*) $T_2 = T_1\, r^{\gamma - 1} = 356 \times 13^{1.4-1} = 993°K$ **Ans.**

cut off ratio $= \frac{v_3}{v_2} = \frac{T_3}{T_2} = \frac{2073}{993} = \beta = 2.08$

(*b*) $T_4 = T_1 \beta^{\gamma} = 356\ (2.08)^{1.4} = 1000°K$ **Ans.**

(*c*) Heat supplied per *kg* of working fluid is $q_S = c_p\ (T_3 - T_2) = 1 \times 0.986\ (1800 - 720)$

$= 1064.88$ KJ

Heat rejected per kg of working fluid is

$q_R = c_v\,(T_4 - T_1)$

$= 1 \times 0.702\ (727 - 83)$

$= 452.08$ KJ **Ans.**

(*d*) Thermal efficiency of the cycle is given by

$$\eta = \frac{\text{Heat supplied} - \text{Heat rejected}}{\text{Heat supplied}}$$

$$= \frac{1064.88 - 452.08}{1064.88} = 57.54\%. \quad \textbf{Ans.}$$

Example 10.20. *In a diesel cycle the lowest temperature and lowest pressure are 300°K and 1 bar respectively. The compression ratio is 16 and heat added per kg of air is 2500 KJ/kg. Determine (a) the temperature and pressure at the end of compression (b) the highest temperature reached in the cycle (c) cycle efficiency, and (d) power output for air flow rate of 0.25 kg/s.*

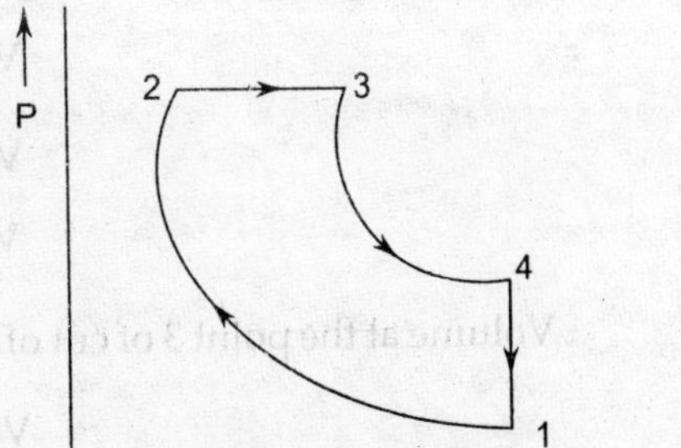

Fig. 10.15

Take $c_p = 1$ KJ/*kg*°K

$c_v = 0.714$ KJ/*kg*°K.

Solution. Given $P_1 = 1$ bar, $T_1 = 300$ °K,

compression ratio $= r = 16$, $q_S = 2500$ KJ/*kg*.

(*a*) $T_2 = T_1 r^{\gamma-1} = 300\,(16)^{1.4-1} = 909°K$ **Ans.**

$P_2 = P_1 r^{\gamma} = 1\,(16)^{1.4} = 48.5$ bar. **Ans.**

(*b*) $q_S = c_p (T_3 - T_2)$

$2500 = 1\,(T_3 - 909) \Rightarrow T_3 = 3409°K$ **Ans.**

$$V_1 = \frac{RT_1}{P_1} = \frac{(C_p - C_v)T_1}{P_1} = \frac{(1-0.714)\times 1000 \times 300}{1\times 10^5}\frac{m^3}{kg}$$

$$= 0.858\ m^3/kg = v_4$$

$$v_2 = \frac{v_1}{r} = \frac{0.858}{16} = 0.0536\ m^3/kg$$

$$v_3 = \frac{T_3}{T_2}.v_2 = \frac{3409}{909}\times 0.0536 = 0.201\ m^3/kg.$$

$$T_4 = \frac{T_3}{\left(\dfrac{v_4}{v_3}\right)^{\gamma-1}} = \frac{3409}{\left(\dfrac{0.858}{0.201}\right)^{1.4-1}} = 1908°K$$

Heat rejected $= c_v (T_4 - T_1)$

$= 0.714\,(1908 - 300)$

$= 1147.9\ KJ/kg.$ **Ans.**

(*c*) Cycle efficiency

$$\eta = \frac{2500 - 1147.9}{2500}\times 100\%$$

$= 54.08\%.$ **Ans.**

(*d*) Power output $= (2500 - 1147.9)\times 0.25$

$= 338$ KW. **Ans.**

Example 10.21. *A C.I engine working on diesel cycle has the following particulars : Stroke = 0.25 m, cylinder bore = 150 mm, clearance volume = 400 cm³. The fuel injection takes place at constant pressure for 5% of the stroke. Find the cycle efficiency. Now the fuel cut off is delayed from 5% to 8% of the stroke. What will be the percentage loss in efficiency if the compression ratio remains the same.*

Solution. Given $l = 0.25\ m,\ d = 150\ mm = 15\ cm$

$V_c = 400\ cm^3 = V_2$

$$V_s = \frac{\pi}{4}d^2 l = \frac{\pi}{4}\times 15^2 \times 25 = 4415\ cc$$

$V_1 = V_c + V_s = 400 + 4415 = 4815\ cc$

$$\text{Volume at the point 3 of cut off} = 400 + \frac{5}{100}\times 4415$$

$V_3 = 620.75\ cc.$

$$\text{Cut off ratio} = \beta = \frac{v_3}{v_2} = \frac{620.75}{400} = 1.552$$

$$r = \frac{4815}{400}\left(=\frac{V_1}{V_2}\right) = 12.04$$

$$\eta = 1 - \frac{\beta^{\gamma-1}}{r^{\gamma-1}\,\gamma(\beta-1)} = 1 - \frac{(1.552)^{1.4}-1}{(12.04)^{0.4}\,1.4\,(1.552-1)}$$

$$= 59.4\% \quad \textbf{Ans.}$$

When the fuel is cut off at 8% of stroke new volume at cut off $= 400 + \frac{8}{100} \times 4415 = 753.2cc$

New $\beta = \frac{753.2}{400} = 1.883$

The new $\eta = 1 - \frac{(1.883)^{1.4}-1}{(12.04)^{0.4}\,1.4\,(1.883-1)}$

$$= 57.4\%$$

% loss in efficiency $= 59.4 - 57.4$

$$= 2\%. \quad \textbf{Ans.}$$

Example 10.22. *A Diesel engine has a compression ratio of 14 and cut-off takes place at 6% of the stroke. Find the air standard efficiency.*

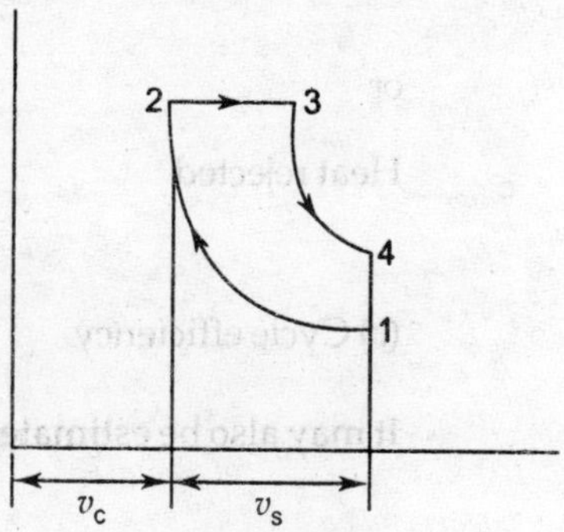

Fig. 10.16

Solution. $r = \frac{V_1}{V_2} = 14$

Cut off volume $= V_3 - V_2 = \frac{6}{100} V_s = \frac{6}{100}(V_1 - V_2)$

$$\frac{V_3}{V_2} - 1 = \frac{6}{100}\left(\frac{V_1}{V_2} - 1\right)$$

or $\frac{V_3}{V_2} = 1 + \frac{6}{100}(14-1) = 1 + \frac{78}{100} = 1.78$

or $\frac{V_3}{V_2} = \beta = 1.78$

$$\eta = 1 - \frac{\beta^{\gamma}-1}{\gamma\, r^{\gamma-1}(\beta-1)} = 1 - \frac{1.78^{1.4}-1}{1.4\,(14)^{0.4}\,(1.78-1)}$$

$$\eta = 1 - \frac{1.78^{1.4}-1}{1.4 \times (14)^{0.4}\,(1.78-1)} = 60.5\%. \quad \textbf{Ans.}$$

Example 10.23. *In an air standard Diesel cycle, the compression ratio is 16, and at the beginning of isentropic compression, the temperature is 15°C and the pressure is 0.1 MPa. Heat is added until the temperature at the end of the constant pressure process is 1480°C. Calculate (a) the cut-off ratio (b) the heat supplied per kg of air, (c) the cycle efficiency, and (d) the mean effective pressure.*

Solution.

$$r = \frac{v_1}{v_2} = 16$$

$$T_1 = 273 + 15 = 288°K$$

$$P_1 = 0.1 \text{ MPa} = 100 \text{ KN}/m^2$$

$$T_3 = 1480 + 273 = 1753 °K$$

$$\frac{T_2}{T_1} = \left(\frac{v_1}{v_2}\right)^{\gamma-1} = 16^{0.4} = 3.03$$

$$T_2 = 288 \times 3.03 = 873°K$$

$$\frac{P_2 v_2}{T_2} = \frac{P_3 v_3}{T_3}, \qquad [P_2 = P_3]$$

(*a*) $$\beta = \frac{v_3}{v_2} = \frac{T_3}{T_2} = \frac{1753}{873} = 2.01 = \text{cut off ratio } \textbf{Ans.}$$

(*b*) Heat supplied $$= q_s = c_p (T_3 - T_2)$$

$$= 1.005 (1753 - 873) = 884.4 \text{ KJ/kg}$$

$$\frac{T_3}{T_4} = \left(\frac{v_4}{v_3}\right)^{\gamma-1} = \left(\frac{v_1}{v_2} \cdot \frac{v_2}{v_3}\right)^{\gamma-1} = \left(\frac{16}{2.01}\right)^{1.4-1} = 2.29$$

or $$T_4 = \frac{T_3}{2.29} = \frac{1753}{2.29} = 766°K$$

Heat rejected $$q_R = c_v (T_4 - T_1) = 0.718 (766 - 288)$$

$$= 343.2 \text{ KJ}/kg.$$

(*c*) Cycle efficiency $$\eta = 1 - \frac{q_R}{q_S} = 1 - \frac{343.2}{884.4} = 61.2\%. \quad \textbf{Ans.}$$

It may also be estimated from the equation

$$\eta = 1 - \frac{\beta^\gamma - 1}{\gamma r^{\gamma-1} (\beta - 1)}$$

$$= 1 - \frac{(2.01)^{1.4} - 1}{1.4 (16)^{0.4} (2.01 - 1)} = 61.2\%$$

$$w_{net} = q_S - q_R = 884.4 - 343.2$$

$$= 541.2 \text{ KJ}/kg$$

w_{net} can also be found out by $w_{net} = q_S \eta$

$$= 884.4 \times 0.612 = 541.2 \text{ KJ}/kg.$$

$$v_1 = \frac{RT_1}{P_1} = \frac{0.287 \times 288}{100} = 0.827 \, m^3/kg$$

$$v_2 = \frac{0.827}{16} = 0.052 \, m^3/kg$$

$$v_s = v_1 - v_2 = 0.827 - 0.052 = 0.775\ m^3/kg$$

(*d*) $$Pm = m.e.p = \frac{w_{net}}{V_s} = \frac{541.2}{0.775} = 698.45\ \text{KPa} \qquad [1N/m^2 = 1Pa]$$

$$Q_R = m\,C_v\,(T_4 - T_1)$$
$$= 0.02308 \times 0.718 \times (874.14 - 320)$$
$$= 9.183\ \text{KJ}$$

$$\left[\text{check } \eta = 1 - \frac{Q_R}{Q_s} = 1 - \frac{9.183}{22.131} = 58.51\%\right]$$

$$W_{net} = Q_S - Q_R = 22.131 - 9.183 = 12.948\ \text{KJ}$$

$$P_m = \frac{W_{net}}{V_s} = \frac{12.948 \times 10^3\ Nm}{19640 \times 10^{-6}\ m^3} = 659.27 \times 10^3\ N/m^2$$

$$IP = \frac{n\,P_m\,\text{LANK}}{60} = \frac{n\,(P_m\,V_s)\text{NK}}{60} = \frac{1 \times 12.948 \times 480 \times \frac{1}{2}}{60}\ \text{KW}$$

$$= \frac{n \times W_{net} \times \text{NK}}{60} = 51.72\ \text{KW} \quad \textbf{Ans.}$$

Example 10.24. *A diesel engine receives air at 0.1 MPa and 300°K in the beginning of compression stroke. The compression ratio is 16. Heat added per kg of air is 1500 KJ. Determine fuel cut off ratio and cycle thermal efficiency. Assume c_p = 1.0 KJ/kg°K.*

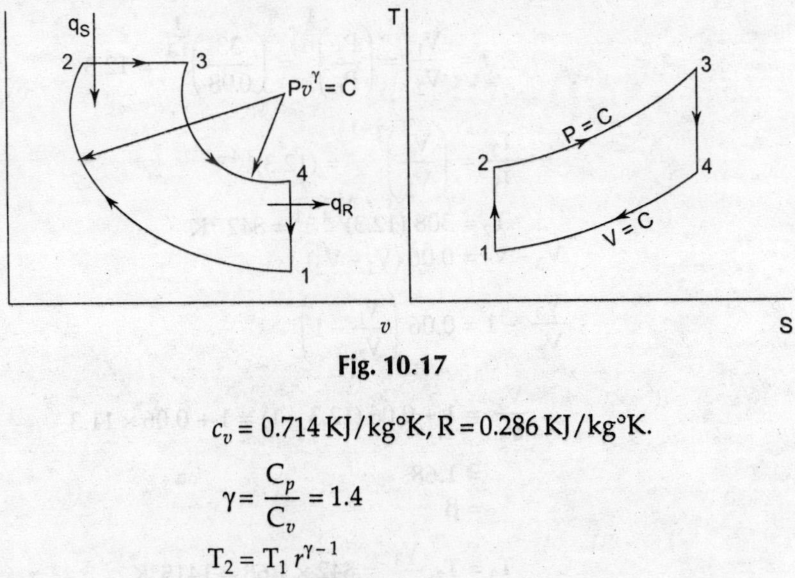

Fig. 10.17

Solution. $$c_v = 0.714\ \text{KJ/kg°K},\ R = 0.286\ \text{KJ/kg°K}.$$

$$\gamma = \frac{C_p}{C_v} = 1.4$$

$$T_2 = T_1\,r^{\gamma-1}$$
$$= 300\,(16)^{1.4-1}$$
$$= 909.43°\text{K}$$

Process 2 – 3 $\dfrac{T_3}{T_2} = \dfrac{v_3}{v_2} = \beta$

Heat added $q_s = c_p (T_3 - T_2)$

$$1500 = 1 (T_3 - 909.43)$$

or $T_3 = 2409.43°K.$

$$\beta = \text{Fuel cut-off ratio} = \frac{T_3}{T_2} = \frac{2409.43}{909.43} = 2.65 \quad \textbf{Ans.}$$

$$\eta_{cycle} = 1 - \frac{\beta^{\gamma} - 1}{\gamma r^{\gamma-1} (\beta - 1)} = 1 - \frac{2.65^{1.4} - 1}{1.4 \times 16^{0.4} (2.65 - 1)}$$

$$= 58.39\% \quad \textbf{Ans.}$$

Example 10.25. *An engine working on ideal diesel cycle operates on 1 kg of standard air with an initial pressure of 0.98 bar, a temperature of 35°C. The pressure at the end of compression is 33 bar and the cut-off is 6% of the stroke. Determine the heat supplied during the cycle and mean effective pressure. Assume c_p = 1.0KJ/kg°K, γ = 1.4.*

Solution.

$$\text{Compression ratio} = \frac{V_1}{V_2} = r$$

$$P_1 = 0.98 \text{ bar}$$

$$T_1 = 273 + 35 = 308°K$$

$$P_2 = 33 \text{ bar}$$

$$V_3 - V_2 = \frac{6}{100} (V_1 - V_2)$$

$$r = \frac{V_1}{V_2} = \left(\frac{P_2}{P_1}\right)^{\frac{1}{r}} = \left(\frac{33}{0.98}\right)^{\frac{1}{1.4}} = 12.3$$

$$\frac{T_2}{T_1} = \left(\frac{V_1}{V_2}\right)^{\gamma-1} = (12.3)^{1.4-1}$$

$$T_2 = 308 (12.3)^{1.4-1} = 842 °K$$

or $V_3 - V_2 = 0.06 (V_1 - V_2)$

or $\dfrac{V_3}{V_2} - 1 = 0.06 \left(\dfrac{V_1}{V_2} - 1\right)$

or $\dfrac{V_3}{V_2} = 1 + 0.06 (12.3 - 1) = 1 + 0.06 \times 11.3$

$$= 1.68$$

$$= \beta$$

$$T_3 = T_2 . \frac{V_3}{V_2} = 842 \times 1.68 = 1415°K$$

Heat added $q_s = c_p (T_3 - T_2) = 1.0(1415 - 842)$

$$= 573 \text{ KJ/kg.}$$

Process 3 – 4 $\dfrac{T_4}{T_3} = \left(\dfrac{V_3}{V_4}\right)^{\gamma-1} = \left(\dfrac{V_3}{V_2}\cdot\dfrac{V_2}{V_4}\right)^{\gamma-1} = \left(\dfrac{V_3}{V_2}\cdot\dfrac{V_2}{V_1}\right)^{\gamma-1}$

or $T_4 = 1415\left(1.68\times\dfrac{1}{12.3}\right)^{1.4-1} = 642^{\circ}\text{K}$

Heat rejected $q_R = c_v (T_4 - T_1)$

$$= 0.714\,(642 - 308) = 238.5 \text{ KJ/kg.}$$

Heat supplied – Heat rejected

$$= (573 - 238.5) \text{ KJ/kg}$$
$$= 334.5 \text{ KJ/kg} = \text{Net workdone}$$

$$P_1 v_1 = RT_1$$

or $v_1 = \dfrac{RT_1}{P_1} = \dfrac{0.286\,(308)\times 10^3}{0.98\times 10^5} = 0.90\ m^3/\text{Kg}$

$$v_2 = \frac{v_1}{r} = \frac{0.90}{12.3} = 0.073\ m^3/\text{Kg}$$

$$Pm = \frac{\text{Net workdone}}{v_s} = \frac{334.5}{v_1 - v_2}\ \frac{\text{KJ/kg}}{m^3/\text{kg}}$$

$$= \frac{334.5}{0.90 - 0.073} = \frac{334.5}{0.827}\ \text{KN}/m^2$$

$$= \frac{334.5}{0.827}\times 10^{-2} \text{ bar}$$

$$= \frac{3.345}{0.827} \text{ bar} = 4.05 \text{ bar.} \quad \textbf{Ans.}$$

Example 10.26. *A 4 cylinder 4 stroke petrol engine develops (IP) 14.7 KW at 1000 rpm. The mean effective pressure is 5.5 bar. Calculate the bore and stroke of engine if the length of stroke is 1.5 times of the bore.*

Solution. Given that $\eta = 4$, power (IP) = 14.7 KW

$$N = 1000\ rpm,\ P_m = 5.5 \text{ bar}$$

$$l = 1.5\,d,\ A = \frac{\pi}{4}d^2$$

$$k = \frac{1}{2}$$

indicated power $(\text{IP}) = \dfrac{n\,P_m\,\text{LANK}}{60}$

$$14.7 = \frac{4\times 5.5\times 100\times 1.5d\times\frac{\pi}{4}d^2\times 1000\times\frac{1}{2}}{60}$$

$$d = 0.0879 \text{ m} = 87.9 \text{ mm.} \quad \textbf{Ans.}$$

$$l = 1.5 \times 0.0879 = 0.1318\, m = 131.8 \text{ mm.} \quad \textbf{Ans.}$$

Example 10.27. *An air standard diesel cycle has a compression ratio of 14. The pressure at the beginning of the compression stroke is 1 bar and the temperature is 300°K. The maximum cycle temperature is 2500°K. Determine the cut-off ratio and thermal efficiency.*

Solution. Given

$$r = 14$$

$$T_1 = 300°K \text{ and } P_1 = 1 \text{ bar}$$

$$T_{max} = T_3 = 2500°K$$

$$\therefore \quad T_3 = T_1\, \beta r^{\gamma-1}$$

or $$2500 = 300\, \beta (14)^{0.4}$$

or $$\beta = 2.9 = \text{cut off ratio} \quad \textbf{Ans.}$$

Thermal efficiency = Air standard efficiency = η

$$\eta = 1 - \frac{\beta^{\gamma} - 1}{\gamma\, r^{\gamma-1}\,(\beta - 1)} = 1 - \frac{(2.9)^{1.4} - 1}{1.4\,(14)^{0.4}\,(2.9 - 1)}$$

$$= 55\%. \quad \textbf{Ans.}$$

Examples Based on Otto & Diesel Mixed

Example 10.28. *A four cylinder, four stroke engine has a displacement volume of 300 cc per cylinder. The compression ratio of the engine is 10 and operates at a speed of 3000 rev/min. The engine is required to develop an output of 40KW at this speed. Calculate the thermal efficiency of the cycle, assuming that the engine operates on the otto cycle and that the pressure and temperature at the inlet conditions are 1 bar and 27°C respectively. If the above engine is operating on diesel cycle and receiving heat at the same rate, calculate the thermal efficiency, and maximum temperature of the cycle. Also compare η of otto with diesel cycle.*

Solution. Refer standard indicated diagram for Otto & Diesel

(*a*) Given for otto

Number of cylinders = 4

Swept volume = 300 cc/cylinder = v_s

$$r = 10$$

$$N = 3000\ rpm$$

$$\text{Power} = 40 \text{ KW}$$

$$P_1 = 1 \text{ bar}$$

$$T_1 = 27°C = 273 + 27 = 300°K$$

$$\frac{v_1}{v_2} = r = 10$$

or $$\frac{v_c + v_s}{v_c} = 10 \text{ or } \frac{v_c + 300}{v_c} = 10,\ v_c = 33.34\ cc$$

$$\eta_{otto} = 1 - \frac{1}{r^{\gamma-1}} = 1 - \frac{1}{10^{1.4-1}} = 60.2\%. \quad \textbf{Ans.}$$

(*b*) for diesel

$$\text{I.P} = \frac{n\,P_m\ \text{LAN K}}{60}\ \text{KW}$$

$$40 = \frac{4 \times P_m \times v_s \times 3000 \times \frac{1}{2}}{60}\ \text{KW} \quad \left[\begin{array}{c} Pm \text{ in KN/m}^2 \\ v_s \text{ in m}^3 \end{array}\right]$$

or $$40 = \frac{4 \times P_m \times 300 \times 3000 \times \frac{1}{2} \times 10^{-6}}{60}$$

or $$P_m = 1.333 \times 10^3\ \text{KN}/m^2$$

$$= 13.33 \text{ bar}$$

$$q_S = c_p (T_3 - T_2), \qquad T_2 = T_1\, r^{\gamma-1}$$

$$q_R = c_v (T_4 - T_1) \qquad = 300 \times 10^{0.4}$$

$$P_1 = 1 \text{ bar} \qquad = 753.57°K$$

$$v_1 = v_c + v_s = 333.34\ cc = v_4$$

$$v_2 = v_c = 33.34\ cc$$

$$P_2 = P_1 \left(\frac{v_1}{v_2}\right)^{\gamma} = P_1 r^{\gamma} = 1 \times (10)^{1.4} = 25.12 \text{ bar}$$

$$P_2 = P_3 = 25.12 \text{ bar.}$$

$$P_m = \frac{q_s - q_R}{v_s}$$

or $$q_S - q_R = P_m\, v_s = 1.333 \times 10^3 \times 300 \times 10^{-6}\ \frac{KN}{m^2} \times m^3$$

$$= 0.4\ \text{KJ/kg}$$

$$c_v (T_3 - T_2) - c_v (T_4 - T_1) = 0.4$$

$$T_3 - T_2 - T_4 - T_1 = \frac{0.4}{0.718} = 0.557$$

$$300\,(\beta)\, 10^{0.4} - 753.57 - 300\, \beta^{1.4} - 300 = 0.557$$

$$\beta\, 10^{0.4} - \beta^{1.4} - 2.5119 - 1 = 0.557$$

$$2.5119\, \beta - \beta^{1.4} - 3.5119 - 0.557 = 0$$

$$\beta^{1.4} - 2.5119\beta + 4.068 = 0$$

$$\beta = 4$$

$$\eta_{Diesel} = 1 - \frac{1}{\gamma r^{\gamma-1}} \frac{(\beta^{\gamma} - 1)}{(\beta - 1)}$$

$$= 43.\%$$

$$T_3 = 300\,\beta\,10^{0.4} = 300 \times 4 \times 10^{0.4}$$

$$= 3014°K. \quad \textbf{Ans.}$$

Example 10.29. *Mathematically prove that the efficiency of Otto cycle is greater than that of Diesel cycle of the same compression ratio.*

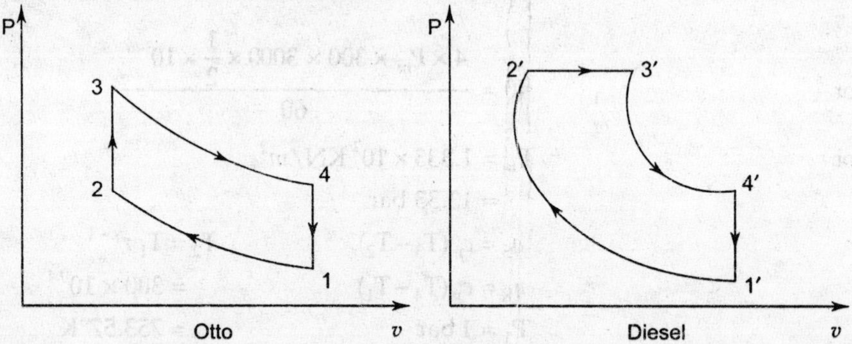

Fig. 10.18

Solution. Given compression ratio of otto = compression ratio of diesel.

$$\frac{v_1}{v_2} = \frac{v_1^1}{v_2^1} = r$$

If $$\beta = \frac{v_3^1}{v_2^1}, \text{ Let expansion ratio} = \frac{v_4^1}{v_3^1} = r_1 = \frac{r}{\beta}$$

$$\beta = \frac{r}{r_1}$$

$$\therefore \eta_{Diesel} = 1 - \frac{\left(\frac{r}{r_1}\right)^{\gamma} - 1}{r^{\gamma-1}\,\gamma\left(\frac{r}{r_1} - 1\right)} \text{ s}$$

From the above we have $\beta > 1$ *i.e.,* $\frac{r}{r_1} > 1$

Let $r_1 = (r - \delta)$ where δ is small quantity.

Then $$\frac{r}{r_1} = \frac{r}{r-\delta} = \left(1 - \frac{\delta}{r}\right)^{-1} = 1 + \frac{\delta}{r} + \frac{\delta^2}{r^2} + \frac{\delta^3}{r^3} + \ldots\ldots \text{ and } \left(\frac{r}{r_1}\right)^{\gamma} = \left(\frac{r}{r-\delta}\right)^{\gamma} = \left(1 - \frac{\delta}{r}\right)^{-\gamma}$$

$$= 1 + \frac{\gamma\delta}{r} + \frac{\gamma(\gamma+1)}{21} \cdot \frac{\delta^2}{r^2} + \ldots\ldots$$

$$\eta_{\text{diesel}} = 1 - \frac{1}{r^{\gamma-1}} \frac{\left[\frac{\gamma\,\delta}{r} + \frac{\gamma(\gamma+1)}{21} \cdot \frac{\delta^2}{r^2} + \ldots\right]}{\gamma\left[\frac{\delta}{r} + \frac{\delta^2}{r^2} + \frac{\delta^3}{r^3} + \ldots\right]}$$

$$\eta_{\text{diesel}} = 1 - \frac{1}{r^{\gamma-1}} \left[\frac{\frac{\delta}{r} + \frac{\gamma+1}{2} \cdot \frac{\delta^2}{r^2} + \ldots}{\frac{\delta}{r} + \frac{\delta^2}{r^2} + \ldots}\right]$$

The ratio inside the bracket is greater than 1 since the co-efficient of terms $\frac{\delta^2}{r^2}$ is greater than 1 in the numerator. It means that something more is subtracted in case of Diesel cycle than in otto cycle because

$$\eta_{\text{otto}} = 1 - \frac{1}{r^{\gamma-1}}.$$

Hence, for the same compressoin ratio $\eta_{\text{otto}} > \eta_{\text{Diesel}}$ **Proved.**

Example 10.30. *An air engine works on the following cycle. Air is taken in at atmospheric pressure of 1.1 kg/cm² and at temperature of 16°C and is compressed adiabatically, the pressure at the end of the stroke being 35 kg/cm² absolute. Heat is taken at constant pressure, the expansion afterwards takes place adiabatically, the ratio of expansion being 5. The air exhausted at the end of stroke, the heat is assumed to be rejected at constant volume. Find the ideal thermal efficiency. Take*

$$c_p = 0.238 \text{ kcal/kg K and } c_v = 0.17 \text{ kcal/kg K.}$$

Solution. The cycle is diesel because

$P = c$ & $v = c$ & $s = c$, $s = c$,

$P_1 = 1.1 \text{ kg/cm}^2 = 1 \text{ bar}$

$T_1 = 16 + 273 = 289°\text{K}.$

$P_2 = 35 \text{ kg/cm}^2 = 34.335 \text{ bar}$

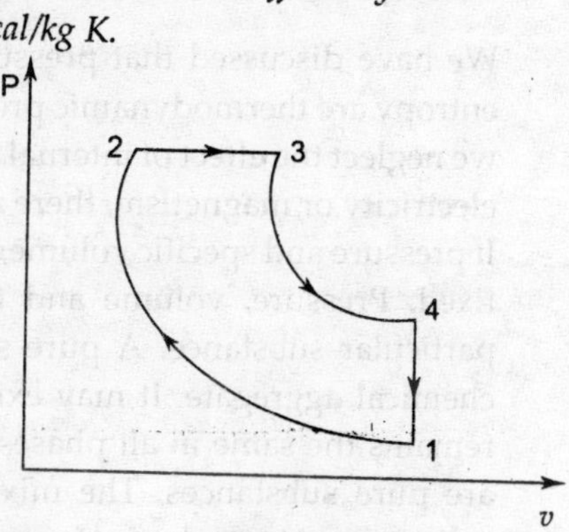

Fig. 10.19

Expansion ratio $= \frac{r}{\beta}$ where

β = cut-off ratio

Process 1 – 2 r = compression ratio

$$\frac{r}{\beta} = 5 = \frac{v_4}{v_3}$$

$$P_1 v_1^\gamma = P_2 v_2^\gamma$$

or

$$r = \frac{v_1}{v_2} = \left(\frac{P_2}{P_1}\right)^{\frac{1}{\gamma}} = \left(\frac{35}{1.1}\right)^{\frac{1}{1.4}} = 11.84$$

$$\beta = \frac{r}{5} = \frac{11.84}{5} = 2.36$$

$$\eta = 1 - \frac{\beta^\gamma - 1}{\gamma r^{\gamma-1}(\beta - 1)} = 1 - \frac{2.36^{1.4} - 1}{1.4\,(11.84)^{0.4}\,(2.36 - 1)}$$

$= 54.5\%$ **Ans.**

Chapter **11**

Properties of Pure Substances

11.1 INTRODUCTION

We have discussed that pressure, temperature, volume, internal energy, enthalpy and entropy are thermodynamic properties as applied to mechanical engineering problems. If we neglect the effect of internal energy due to gravity, observable motion, surface tension, electricity or magnetism, there are only two independent properties for a pure substance. If pressure and specific volume, for example, are fixed then all the other properties become fixed. Pressure, volume and temperature are often used to describe the states of a particular substance. A pure substance is one that is homogeneous and invariable in chemical aggregate. It may exist in one or more phases, but the chemical composition remains the same in all phases. Liquid water, solid water (*ice*) or water vapour (steam) are pure substances. The mixtures of ice and water, and water and steam are pure substance. Atmospheric air is a pure substance so long it remains a gas but if there is a change of phase and we have a mixture of gaseous and liquid air, it cannot be considered as a pure substance since its phases are having different chemical compositions.

11.2 HEATING OF ICE

Fig. 11.1 refers unit mass of ice at – 20°C under one atmospheric pressure. Let us increase the temperature of the ice under constant pressure and observe. We see the following (Refer Fig. 11.2)

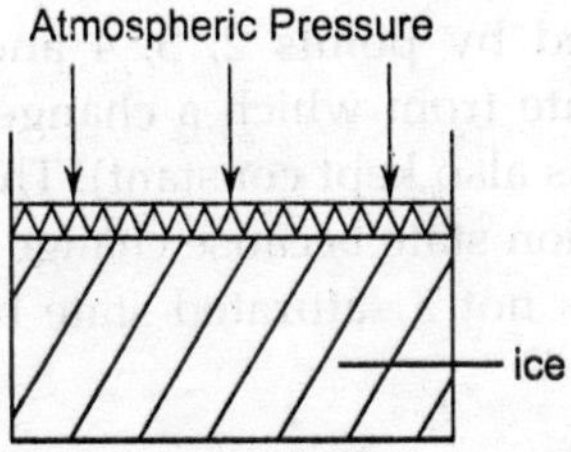

Fig. 11.1

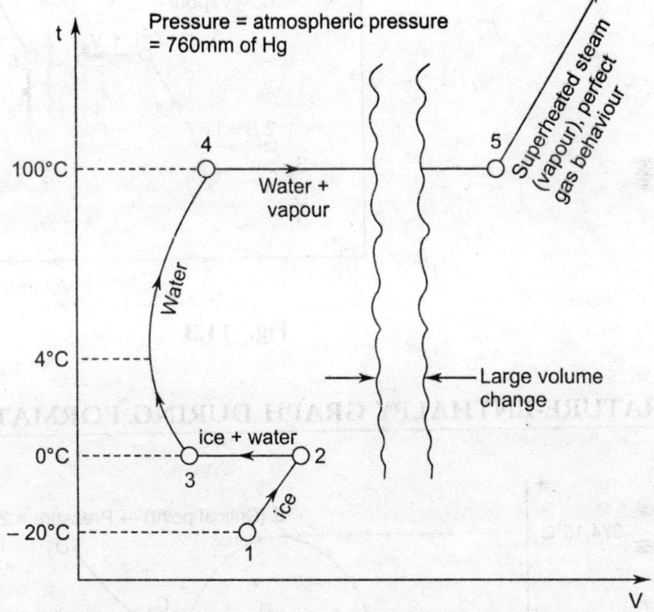

Fig. 11.2

Process 1-2 : The temperature of ice increases from – 20°C to 0°C and there is a small change of volume. The ice is now at 0°C.

Process 2-3 : The solid ice at 0°C starts melting to liquid without any change in temperature. There is a decrease in volume.

Process 3-4 : The temperature of the liquid water at 0°C increases to 100°C. The volume decreases to a minimum at 4°C and then increases upto point 4.

Process 4-5 : The liquid water at 100°C starts boiling to a vapour or gas at constant temperature with a large increase in volume.

Process 5 onwards : Both the temperature and volume of the vapour increases.

Note. The decrease in volume during melting is a peculiarity of water.

11.3 HEATING OF PURE SUBSTANCES OTHER THAN ICE

In general substances show a continuous increase of volume with increase in temperature (Fig. 11.3). The path is reversed when the substance is cooled slowly.

The states indicated by points 2, 3, 4 and 5 are known as saturation states. A saturation state is a state from which a change of phase may occur without changing temperature (pressure is also kept constant). Thus point 2 indicates saturated solid state. The point 3 is a saturation state because change of phase from liquid to solid may occur upon cooling. State 1 is not a saturated state because temperature will change before melting occurs.

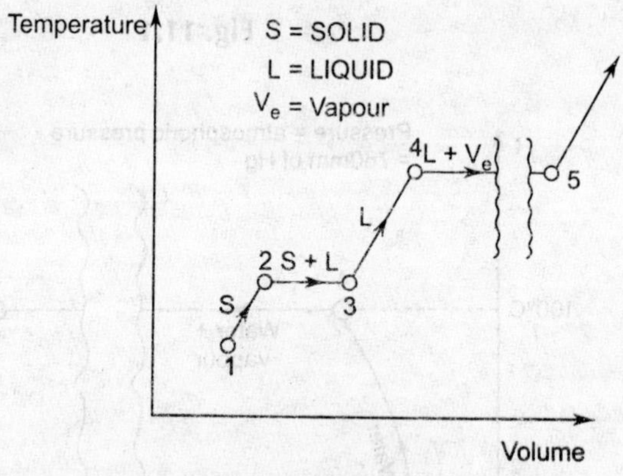

Fig. 11.3

11.4 TEMPERATURE-ENTHALPY GRAPH DURING FORMATION OF STEAM

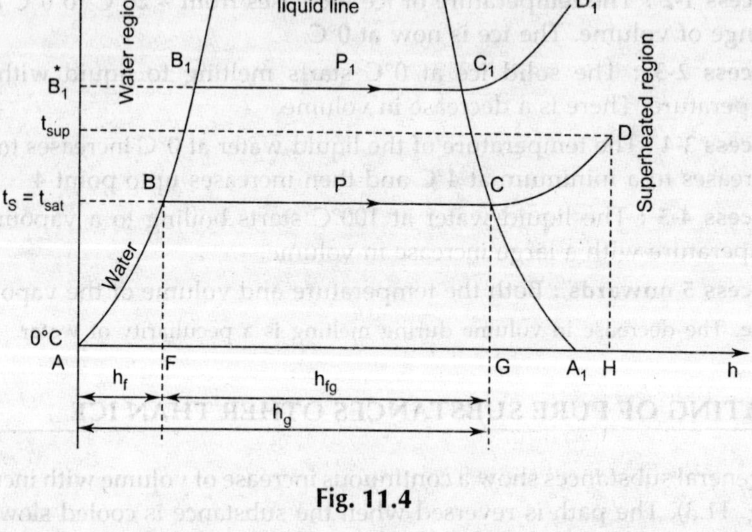

Fig. 11.4

The water at A is having temperature of 0°C. The different stages are as follows:

(*a*) **Induction stage :** AB shows the heating of water upto boiling temperature. The heat required to raise the temperature of water (1 kg mass) from 0°C to boiling point (B) is known as sensible heat of water. It is denoted as h_f.

$$\therefore \quad \text{Sensible heat} = \text{mass of water} \times \text{specific heat} \times \text{rise in temperature}$$
$$= 1 \times 4.2\,[t_s + 273 - 273] = 4.2\,t_s.$$

where specific heat of water is taken as 4.2 KJ/kg.

(*b*) **Warming stage :** BC shows the chage of state from liquid to steam. The heat absorbed during this stage is FG, known as latent heat of vaporisation (h_{fg}).

(*c*) **Evaporation stage :** CD shows the superheating process. The heat absorbed during this stage is GH, known as heat of superheat. Line AH represents the total heat required to go upto superheated point D.

11.4.1 Effect of Change of Pressure

If we increase the pressure to P_1 (bar), the boiling temperature also increases. The point *B_1 represents the boiling temperature or saturating temperature at P_1 bar. The point C_1 is the point on dry saturated steam. Line C_1D_1 shows constant pressure process in which the steam is superheated (after C_1 point). In a similar manner a family of curves may be drawn for different pressures. The line A–B–B_1–B_2 is called saturated liquid line which forms boundary line between water and steam. Similarly C_2–C_1–C–A_1 is called dry saturated steam line which forms boundary line between wet and superheated steam. As the pressure increases the latent heat of vaporisation decreases (saturation temperature increases). It becomes zero at E where liquid and dry steam lines meet. This point E is called critical point. For steam, critical temperature is 374.15°C and critical pressure is 221.2 bar.

Total heat (enthalpy) of dry saturated steam (point C) is $h_g = h_f + h_{fg}$.

where h_g = enthalpy of 1 kg of dry saturated steam

h_f = sensible heat of 1kg of water

h_{fg} = enthalpy of vaporisation of 1 kg of dry saturated steam = Latent heat of steam.

The enthalpy of heating of water upto saturation temperature

$$h_f = c_{pf}\,(T_{sat} - 273.16)$$

where c_{pf} = 4.18 KJ/kg °k.

(*a*) **Wet steam :** If

m_g = mass of actual dry steam

m_f = mass of water in suspension

m = mass of wet steam

$= m_g + m_f.$

(*b*) **Dryness fraction :** (x)

$$x = \frac{m_g}{m_g + m_f}.$$

x is also called the quality of wet steam

for $x = 0$, it is saturated liquid

for $x = 1$, it is dry and saturated steam.

$$0 \le x \le 1.$$

(c) **Internal energy :** $h_f = u_f + Pv_f$

f stands for fluid (water)

$$u_f = h_f - Pv_f$$

for dry steam $u_g = h_g - Pv_g$

g stands for gas (vapour).

(d) sp. enthalpy of wet steam:

$$h = h_f + xh_{fg}$$

(e) sp. enthalpy of superheated steam:

$$h_{sup} = h_f + h_{fg} + c_{pg}\,(t_{sup} - t_{sat})$$

c_{pg} for steam = 2.1 KJ/kg°k.

$t_{sup} - t_{sat}$ = Degree of superheat.

(f) **Specific volume:** Specific volume of wet steam

$$v = v_f + x\,(v_g - v_f)$$
$$= xv_g + (1 - x)\,v_f$$

Charles law when $P = C$

$$\frac{v_{\text{sup}}}{T_{\text{sup}}} = \frac{v_g}{T_{\text{sat}}}$$

(g) **Specific entropy :** specific entropy of wet steam

$$S = S_f + x\,(S_g - S_f)$$

(h) Workdone during evaporation

$$= P\,(v_g - v_f)$$
$$= Pv_g \text{ as } v_f = \text{negligible}$$

(i) Workdone during evaporation of a wet state

$$= Px\,v_g.$$

(j) workdone by superheated steam

$$\approx Pv_{sup}.$$

(k) specific entropy of steam:

$$s_f = \int_{T_1}^{T_2} \frac{đQ}{T} = \int_{T_1}^{T_2} \frac{c_p\,dT}{T}$$

The steam is produced at constant pressure. The entropy is assumed to be zero at 0°C.

Change in specific entropy $= ds = \dfrac{đQ}{T}$.

(i) The heat supplied for heating 1 kg of water at constant pressure is given by

$$đ\theta = c_{pf}\,dT$$

or $$ds = \frac{c_{pf}\,dT}{T}.$$

Integrating $$\int_0^{S_f} ds = c_{pf}\int_{273}^{T_{sat}} \frac{dT}{T} = c_{pf}\,ln\,\frac{T_{sat}}{273}$$

or
$$s_f = c_{pf}\, ln\, \frac{T_{sat}}{273}.$$

(*ii*) The change in entropy during evaporation is given by
$$s_{fg} = \frac{h_{fg}}{T_{sat}}.$$

(*iii*) Entropy for dry saturated steam is given by
$$s_g = s_f + s_{fg}$$
$$= c_{pf}\, ln\, \frac{T_{sat}}{273} + \frac{h_{fg}}{T_{sat}}.$$

(*iv*) Entropy during superheating of steam is given by
$$s_{gj} = \int_{T_{sat}}^{T_{sup}} c_{pg} \frac{dT}{T} = c_{pg}\, ln\, \frac{T_{sup}}{T_{sat}}$$

(*v*) Entropy of superheated steam is given by
$$s_{sup} = s_g + s_{gj}$$
$$= c_{pf}\, ln\, \frac{T_{sat}}{273} + \frac{h_{fg}}{T_{sat}} + c_{pg}\, \ln \frac{T_{sup}}{T_{sat}}$$

where $c_{pg} = s_p$ heat of superheated steam
$= 2.1\ KJ/kg°K$

and $c_{pf} = 4.18\ KJ/kg°K.$

11.5 TEMPERATURE AND SPECIFIC ENTROPY DIAGRAM FOR STEAM

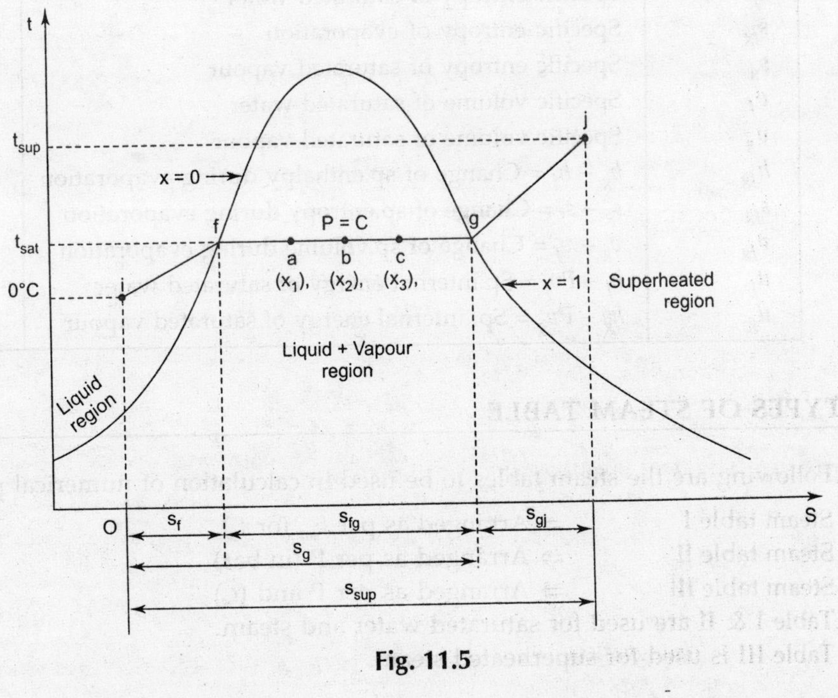

Fig. 11.5

Fig. 11.5 shows temperature entropy diagram for 1 kg of water. At 0°C the entropy is zero (arbitrarily chosen). The steam is generated at constant pressure. *if* shows the heating at constant pressure. The area under the curve '*if*' represents heat transferred to water during the process of heating from 0°C to saturation temperature of water. The change of entropy is S_f during '*if*'. The entropy change S_{fg} occurs during change of phase brought about by the heat flow during evaporation. All datas regarding steam is available at 'steam table' - (interpolation may be done to get exact value).

$$\frac{fa}{fg} = x_1, \frac{fb}{fg} = x_2 \text{ and } \frac{fc}{fg} = x_3$$

There is only one saturation temperature for each pressure.

Six thermodynamic properties of steam are :

(*i*) Pressure(P) (*ii*) sp.volume (v) (*iii*) Temperature (T)
(*iv*) sp. internal energy (u) (*v*) sp. enthalpy (h) (*vi*) specific entropy (s).

11.6 THERMODYNAMIC PROPERTIES AND THEIR UNITS

Following table describes some thermodynamic properties and their units.

Notation	*Description*	*Unit*
P	Absolute pressure	KP_a, or bar
t_s	Saturation temperature	°C
h_f	Specific enthalpy of saturated water	KJ/kg
h_{fg}	Latent heat of evaporation	KJ/kg
h_g	Specific enthalpy of saturated vapour	KJ/kg
s_f	Specific entropy of saturated water	KJ/kg°K
s_{fg}	Specific entropy of evaporation	KJ/kg°K
s_g	Specific entropy of saturated vapour	KJ/kg°K
v_f	Specific volume of saturated water	m^3/kg
v_g	Specific volume of saturated vapour	m^3/kg
h_{fg}	$h_g - h_f$ = Change of sp.enthalpy during evaporation	KJ/kg
s_{fg}	$s_g - s_f$ = Change of sp.entropy during evaporation	KJ/kg
v_{fg}	$v_g - v_f$ = Change of sp.volume during evaporation	m^3/kg
u_f	$h_f - Pv_f$ = Sp.internal energy of saturated water	KJ/kg
u_g	$h_g - Pv_g$ = Sp. internal energy of saturated vapour	KJ/kg

11.7 TYPES OF STEAM TABLE

Following are the steam tables to be used in calculation of numerical problems.

Steam table I ⇒ Arranged as per t_{sat} (or t_s).
Steam table II ⇒ Arranged as per P (in bar)
Steam table III ⇒ Arranged as per P and (t_s)
Table I & II are used for saturated water and steam.
Table III is used for superheated steam.

11.8 MOLLIER DIAGRAM (h-s AXIS)

The following lines are given Mollier Diagram:
(i) Constant h - lines (horizontal)
(ii) Constant s - lines (vertical lines)
(iii) Constant P - lines
(iv) Constant v - lines
(v) Constant t - lines
(vi) Constant x - lines.

11.8.1 Isentropic Process (s = c)

Point 1 : Intersection of P_1 & t_1 (initial point located).
For point 2 : As s = constant, we draw vertical line from 1 to intersect P_2 (given) at 2.

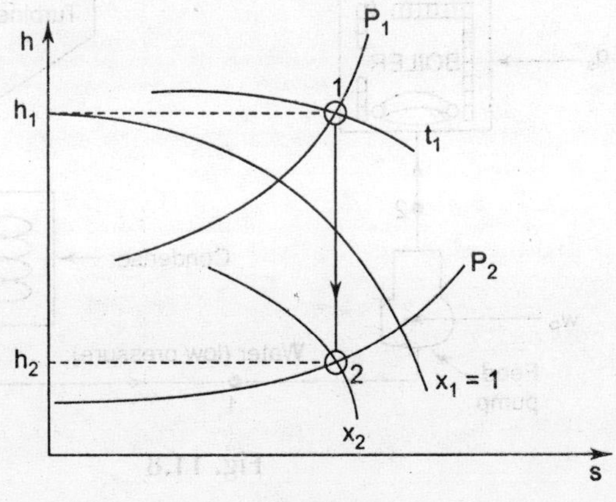

Fig. 11.6

It is observed (as an example) the steam is wet at 2, with dryness fraction of x_2 ($< x_1$). The enthalpy drop $h_1 - h_2$ may be read out from the chart.

11.8.2 Throttling Process (h = const.)

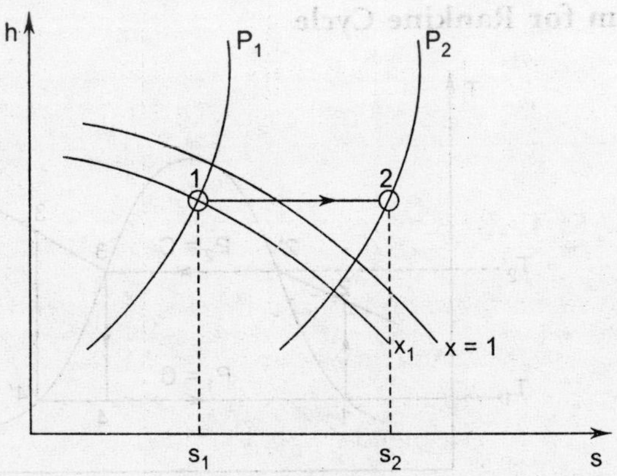

Fig. 11.7

Point 1 : Intersection of P_1 and x_1 (given).
For point 2 : Draw horizontal line to meet P_2 at 2.
For example, it is seen from the diagram that the steam at point 2 is superheated.

11.9 STEAM POWER PLANT (SIMPLE RANKINE CYCLE)

In steam power plant Rankine cycle is used. Fig. 11.8 shows schematic diagram for steam power plant. It consists of four steady state steady flow processes.

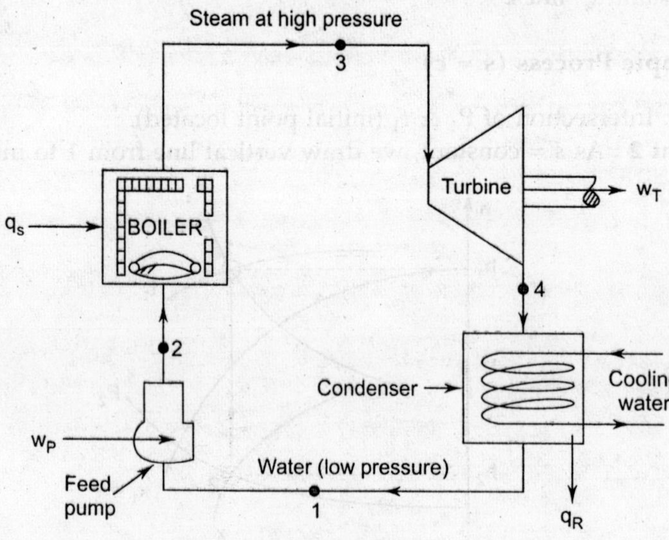

Fig. 11.8

Process	Description
1 – 2	Reversible adiabatic pumping
2 – 3	Heat addition at constant pressure
3 – 4	Reversible adiabatic expansion in turbine
4 – 1	Heat rejection at constant pressure in the condenser.

11.9.1 T–S Diagram for Rankine Cycle

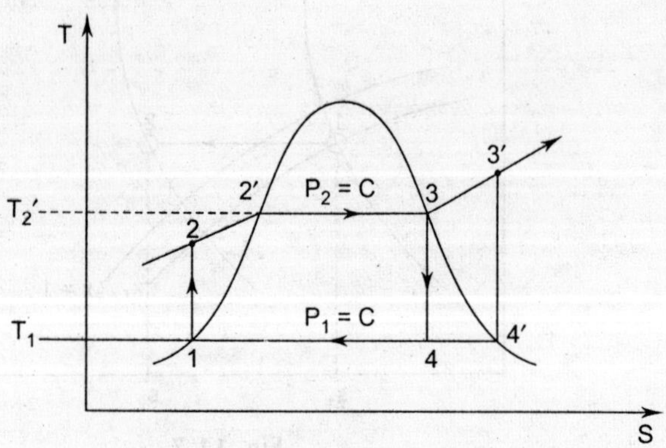

Fig. 11.9

The process is 1–2–2′–3–4–1 cyclic.

In case the boiler supplies superheated steam, the cycle will be 1–2–2′–3–3′–4′–1.

11.9.2 P-*v* Diagram for Rankine Cycle

Here 1–2 is not vertical line it is curve. Similarly 3 to 4 and 3′ to 4′.
2–2′–3–3′ is constant pressure one line. Also 1–4–4′ constant pressure one line.

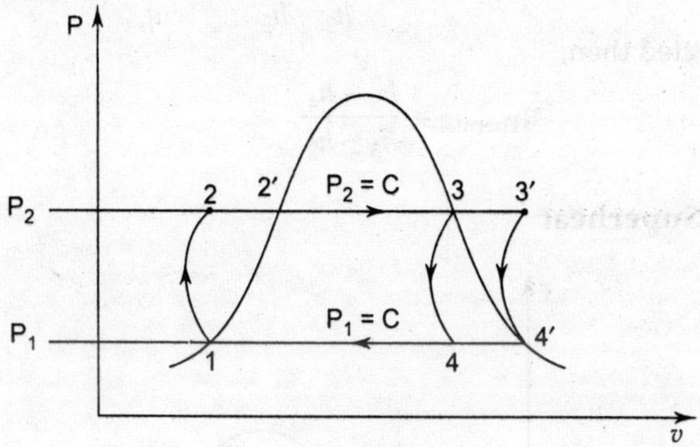

Fig. 11.10

11.9.3 h-s Diagram for Rankine Cycle

1–2 vertical line, 3–4 and 3′–4 vertical lines, 2–2′–3–3′ on curve line also 4′–4–1 on curve line

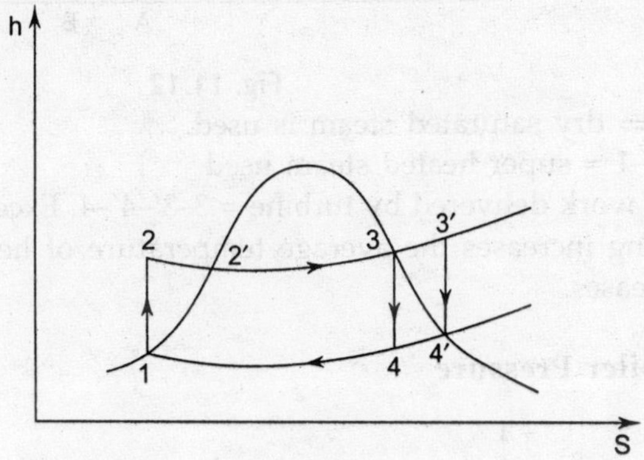

Fig. 11.11

11.9.4 Thermal Efficiency Calculations

For Process 1–2 $\quad w_p = h_2 - h_1$

For Process 2–3 $\quad q_s = h_3 - h_{2'}$ $\quad [\text{đ}q = dh + \text{đ}w, \text{đ}w = -v\int dP = 0 \text{ as } P = c]$

For Process 3–4 $\quad w_T = h_3 - h_{4'}$ $\quad [\therefore\ \text{đ}q = dh]$

For Process 4–1 $\quad q_R = h_4 - h_1$

For Net work

$$= w_T - w_P$$

$$= (h_3 - h_4) - (h_2 - h_1)$$

$$\text{Heat supply} = q_s = h_3 - h_2$$

$$\therefore \quad \eta_{\text{Thermal}} = \frac{\text{Net work}}{\text{heat supply}} = \frac{(h_3 - h_4) - (h_2 - h_1)}{h_3 - h_2}$$

$$= \frac{(h_3 - h_2) - (h_4 - h_1)}{h_3 - h_2}$$

$$= 1 - \frac{h_4 - h_1}{h_3 - h_2} = 1 - \frac{q_R}{q_s} \text{ (obvious)}$$

If w_p is neglected then,

$$\eta_{\text{Thermal}} = \frac{h_3 - h_4}{h_3 - h_2}.$$

11.9.5 Effect of Superheat

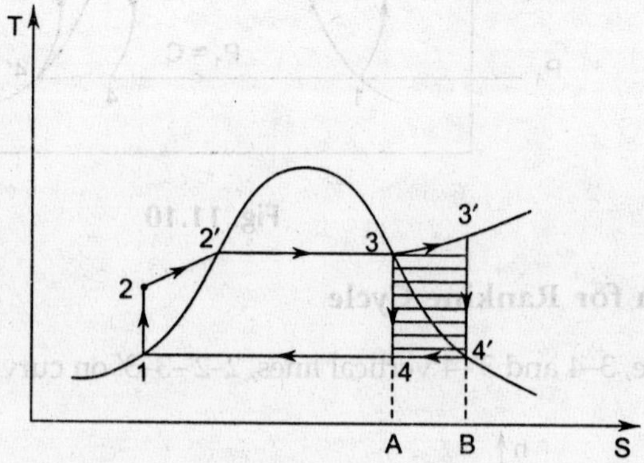

Fig. 11.12

1–2–2′–3–4–1 ⇒ dry saturated steam is used.
1–2–2′–3–3′–4′–1 = super heated steam used.
There is more work delivered by turbine = 3–3′–4′–4. Excess heat supply = 3–3′–B–A.
As superheating increases the average temperature of heat addition to the cycle, the efficiency increases.

11.9.6 Effect of Boiler Pressure

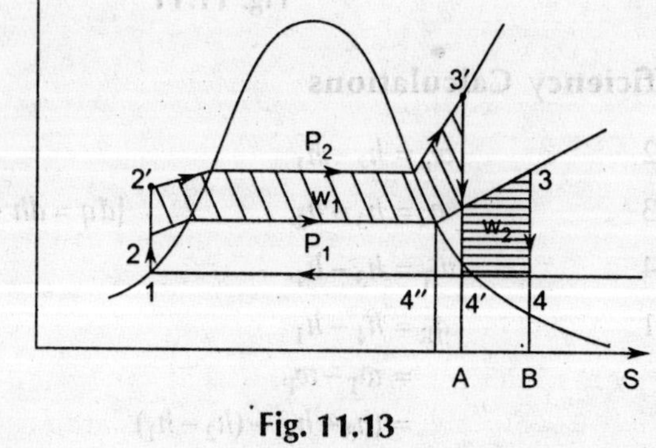

Fig. 11.13

$P_2 > P_1 \Rightarrow$ Increase in boiler pressure to P_2, condenser pressure same at $w_1 = w_2$ (area of the shaded portion). But heat rejection is decreased by 4–B–A–4′. Thus efficiency increases.

11.9.7 Effect of Condenser Pressure

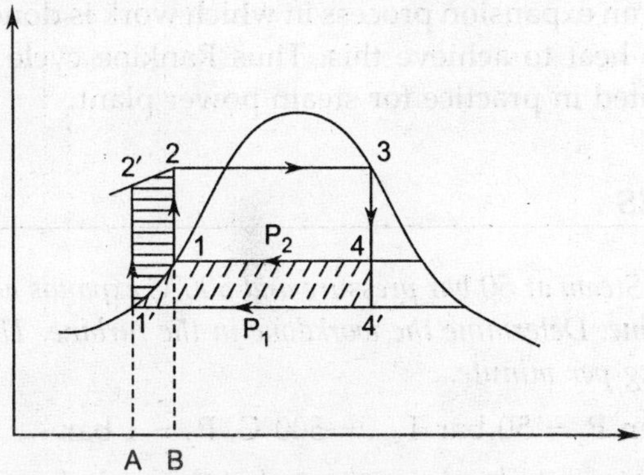

Fig. 11.14

$P_1 < P_2 \Rightarrow$ Decrease in condenser pressure.
Increase in work $\Rightarrow$ 1–4–4′–1′–2′–2–1.
Increase in heat supply to steam $\Rightarrow$ A–2′–2–B. Increase in work > Increase in heat supply
Thus η increases.

11.9.8 Comparison between Carnot Cycle and Rankine Cycle

Carnot cycle
$\Rightarrow$ 1′–2′–3–4–1′
Rankine cycle
$\Rightarrow$ 1–2–2′–3–3′–4′–1

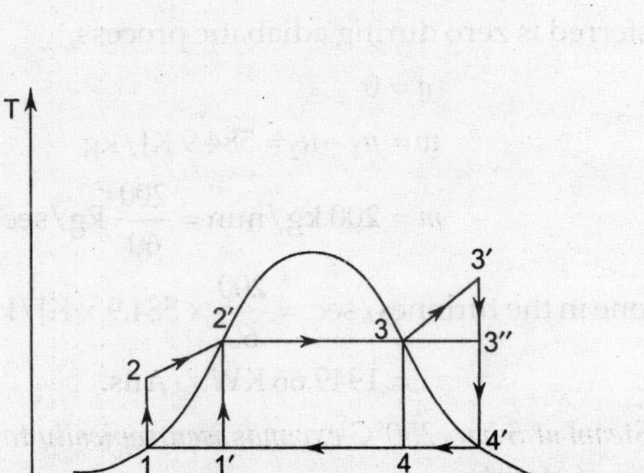

Fig. 11.15

Rankine cycle is used for steam, Carnot cycle is not used for steam – this may be explained by following.

(*a*) **Pumping process :** State 1′ is a mixture of liquid and vapour. We face great difficulty to encounter in manufacturing a pump that will handle the mixture of liquid +vapour at 1′ and deliver saturated liquid at 2′. It is easier to condense the vapour completely and handle only liquid in the pump. The Rankine cycle is based on this fact.

(*b*) **Superheating the vapour:** 3–3′ is a constant pressure process where the vapour is superheated. In Carnot the operation of heat transfer is carried out at constant temperature, and therefore the vapour is superheated in process 3–3″. However the pressure drops during the process which results in heat transfer to the vapour as it undergoes an expansion process in which work is done. There is a practical difficulty to transfer heat to achieve this. Thus Rankine cycle is the ideal cycle that can be approximated in practice for steam power plant.

SOLVED EXAMPLES

Example 11.1. *Steam at 50 bar pressure and 500°C expands adiabatically to a pressure of 1 bar in a steam turbine. Determine the workdone in the turbine. The rate of steam flow through the turbine is 200 kg per minute.*

Solution. Given P_1 = 50 bar T_{sup} = 500°C, P_2 = 1 bar.

Assuming the steam to be dry saturated at the end of expansion. From steam tables for superheated steam corresponding to a pressure of 50 bar and 500°C, we find that

$$u_1 = 3091 \text{ KJ/kg}$$

and from steam tables for saturated steam corresponding to a pressure of 1 bar, we find that

$$u_2 = 2506.1 \text{ KJ/kg}$$

$$\text{Change of internal energy} = u_1 - u_2 = 584.9 \text{ KJ/kg.}$$

Applying 1st law of thermodynamics

$$q = w + (u_2 - u_1)$$

Since heat transferred is zero during adiabatic process,

$$q = 0$$

$$\therefore \quad w = u_1 - u_2 = 584.9 \text{ KJ/kg.}$$

Since $$m = 200 \text{ kg/min} = \frac{200}{60} \text{ kg/sec.}$$

$$\text{(power) workdone in the turbines/sec.} = \frac{200}{60} \times 584.9 \times \text{KJ/kg}$$

$$= 1949.66 \text{ KW.}$$ **Ans.**

Example 11.2. *Steam at 5 bar, 250°C expands isentropically to a pressure of 0.7 bar. Determine the final condition of steam.*

Solution. Let x_2 = Final dryness fraction of steam. The problem may be solved either by steam tables or by Mollier diagram.

1. By Steam Tables : In an isentropic process, entropy remains constant.

$$s_1 = s_2$$

$$\downarrow \quad \downarrow$$

$$s_{sup} = (s_f + x_2 \, s_{fg})$$

From steam tables of superheated steam, corresponding to a pressure of 5 bar and 250°C, we find that

$$s_{sup} = 7.271 \text{ KJ/kg K}$$

From dry and saturated steam table at 0.7 bar pressure.

$$s_f = 1.192 \text{ KJ/kg K}$$

$$s_{fg} = 6.288 \text{ KJ/kg K}$$

$$\therefore \quad s_{sup} = s_f + x_2 \, s_{fg}$$

$$7.271 = 1.192 + x_2 \, (6.288)$$

or $\quad x_2 = 0.966$ **Ans.**

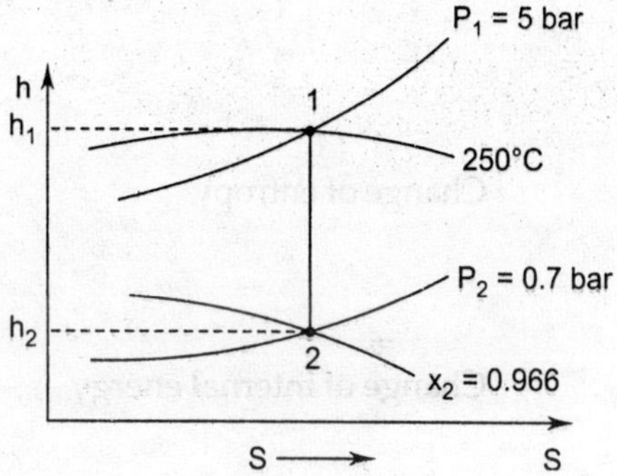

Fig. 11.16

2. **By Mollier diagram:** The final condition of steam (x_2) may be obtained by using Mollier diagram as shown in Fig. 11.16. A vertical line is drawn from point 1 at an initial condition of steam *i.e.*, pressure 5 bar and temperature 250°C, upto the pressure line of 0.7 bar. The dryness fraction is read at the point of intersection (*i.e.*, 2) from which $x_2 = 0.966$.

Example 11.3. *A cylinder contains steam at 1 bar and temperature 150°C. The steam is compressed reversibly and isothermally (T = C) to a state where the specific volume is 0.28 m³/kg. Find the change of internal energy, change of entropy and workdone per kg of steam.*

Solution. Given data, isothermal process

$$P_1 = 1 \text{ bar} = 100 \text{ KN/m}^2$$

$$t_1 = t_2 = 150°C$$

superheated steam table at

$$P_1 = 1 \text{ bar and } 150\ °C \text{ we get,}$$

$$vg_1 = 1.936 \text{ m}^3/\text{kg}$$

$$hg_1 = 2776.4 \text{ KJ/kg}$$

$$sg_1 = 7.613 \text{ KJ/kg K}$$

$$u_1 = u_{g1} = 2582.8 \text{ KJ/kg or}$$

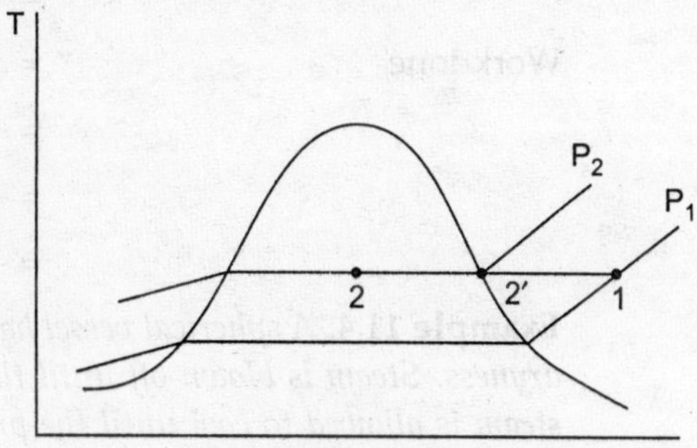

Fig. 11.17

$$\begin{bmatrix} h_{g1} = & u_{g1} + P_1 vg_1 \\ 2776.4 = & u_{g1} + 100 \times 1.936 \end{bmatrix}$$

Since the process is isothermal, the saturation pressure, corresponding to 150°C is 4.8 bar (P_2) in steam table.

Hence

$$P_2 = 4.8 \text{ bar} \qquad h_{f2} = 633.6 \text{ KJ/kg}$$

$$vg_2 = 0.39 \qquad h_{fg2} = 2113.2 \text{ KJ/kg}$$

$$v_2 = x_2 \, vg_2 \qquad s_{f2} = 1.845 \text{ KJ/kg}$$

$$0.28 = x_2 \times 0.39 \qquad s_{fg2} = 4.990 \text{ KJ/kg}$$

$$x_2 = 0.718$$

$$h_2 = h_{f2} + x_2 h_{fg2}$$
$$= 633.6 + 0.718 \times 2113.2$$
$$= 2150.8 \text{ KJ/kg.}$$
$$u_2 = h_2 - P_2 v_2$$
$$= 2150.8 - 4.8 \times 100 \times 0.28$$
$$= 2016.4 \text{ KJ/kg}$$
$$s_2 = s_{f2} + x_2 s_{fg2}$$
$$= 1.845 + 0.718 \times 4.990$$
$$= 5.428 \text{ KJ/kg K}$$

Change of entropy $= s_2 - s_1$
$$= 5.428 - 7.613$$
$$= -2.185 \text{ KJ/kg K.}$$ **Ans.**

Change of internal energy $= u_2 - u_1$
$$= 2016.4 - 2582.8 \text{ KJ/kg}$$
$$= -566.4 \text{ KJ/kg}$$

Heat transferred
$$q = T(s_2 - s_1)$$
$$= (273 + 150)(-2.185)$$
$$= -924.3 \text{ KJ/kg}$$

Workdone $= q - (u_2 - u_1)$
$$= -924.3 - (-566.4)$$
$$= -924.3 + 566.3$$
$$= -358 \text{ KJ/kg.}$$ **Ans.**

Example 11.4. *A spherical vessel having a capacity of 0.85 m^3 contains steam at 12 bar and 0.95 dryness. Steam is blown off until the pressure drops to 6 bar. The valve is then closed and the steam is allowed to cool until the pressure falls to 4 bar. Assuming that the enthalpy per kg of steam in the vessel remains constant during blowing off period, determine,*

(i) The quantity of steam blown off.
(ii) The quality of steam in the vessel after cooling.
(iii) The heat lost by steam per kg during cooling.

Solution. Given data

$$P_1 = 12 \text{ bar}$$
$$P_2 = 6 \text{ bar}$$
$$V = 0.85 \text{ m}^3$$
$$x_1 = 0.95$$
$$P_3 = 4 \text{ bar}$$

Process 1–2. Constant enthalpy (adiabatic)

$$h_1 = h_2$$

From dry and saturated steam table at 12 bar pressure

$$h_{f1} = 798.6 \text{ KJ/kg}$$

$$h_{fg1} = 1986.2 \text{ KJ/kg}$$

$$vg_1 = 0.163 \text{ m}^3\text{/kg}$$

Capacity of vessel $\quad V = m_1 x_1 Vg_1$

$$0.85 = m_1 \times 0.95 \times 0.163$$

or $\quad m_1 = 5.489 \text{ kg}$

$$h_1 = h_{f1} + x_1 h_{gf1}$$

$$= 798.6 + 0.95 \times 1986.2$$

$$= 2685.49 \text{ KJ/kg}$$

$$h_2 = h_1 = 2685.49 \text{ KJ/kg.}$$

From dry and saturated steam table at 6 bar pressure

$$h_{f2} = 670.6 \text{ KJ/kg}$$

$$h_{fg2} = 2086.3 \text{ KJ/kg}$$

$$vg_2 = 0.316 \text{ m}^3\text{/kg}$$

$$h_2 = h_{f2} + x_2 \times h_{fg2}$$

$$2685.49 = 670.6 + x_2 \times 2086.3$$

or $\quad x_2 = 0.965$

$$v_2 = X_2\, vg_2 = 0.965 \times 0.316$$

$$= 0.305 \text{ m}^3\text{/kg.}$$

$$V = m_2 v_2$$

$$0.85 = m_2 \times 0.305$$

or $\quad m_2 = 2.785 \text{ kg}$

Quantity of steam blown off $= m_1 - m_2$

$$= 5.489 - 2.785$$

$$= 2.704 \text{ kg.} \quad \textbf{Ans.}$$

Process 2–3. Constant volume ($v_2 = v_3 = 0.305 \text{ m}^3\text{/kg}$)

From dry and saturated steam table at 4 bar pressure

$$h_{f3} = 604.7 \text{ KJ/kg} \qquad v_3 = x_3 vg_3$$

$$h_{fg3} = 2133.8 \text{ KJ/kg} \qquad 0.305 = x_3 \times 0.463$$

$$v_{g3} = 0.463 \, m^3\text{/kg} \qquad \text{or } x_3 = 0.658$$

Quality of steam in the vessel after cooling

$$= 0.658$$

$$h_3 = h_{f3} + x_3\, h_{fg3}$$

$$= 604.7 + 0.658 \times 2133.8$$

$$= 2008.74 \text{ KJ/kg}$$

Heat lost by steam during cooling

$$= (h_2 - h_3) - v_3 (P_2 - P_3)$$

$$= (2685.49 - 2008.74) - 0.305\,(6 - 4)100$$

$$= 615.75 \text{ KJ/kg } \textbf{Ans.}$$

Example 11.5. *0.05 kg of steam at 15 bar is contained in a rigid vessel of volume 0.0076 m^3. What is the temperature of the steam? If the vessel is cooled, at what temperature will the steam be just dry saturated? Cooling is continued until the pressure in the vessel is 11 bar. Calculate the final dryness fraction of the steam.*

Solution. Given data $m = 0.05$ kg

$P_1 = 15$ bar

$V = 0.0076 \text{ m}^3$

$P_2 = 11$ bar

The steam is superheated

$$V = m\, v_1$$

$$0.0076 = 0.05\, v_1$$

or

$$v_1 = 0.152\, m^3/\text{kg}.$$

From superheated steam table, corresponding to a steam press of 15 bar and volume 0.152 m^3/kg, the temperature of steam is 250°C. [$v_g = 0.152$ at 15 bar].

Process 1-2. Constant volume cooling ($v_1 = v_2$) from dry and saturated steam table, corresponding to the pressure of 15 bar the saturation temperature of steam is 191.6°C. From dry and saturated steam table corresponding to the pressure of 11 bar we get.

$$v_{g2} = 0.178 \text{ m}^3/\text{kg}$$

$$v_1 = v_2 = x_2 \times v_{g2}$$

$$0.152 = x_2 \times 0.178$$

or

$$x_2 = 0.854.$$

Thus the steam will be just dry saturated at the temperature of 191.6°C. Final dryness fraction of steam = 0.854. **Ans.**

Example 11.6. *Find the internal energy of 1 kg of steam at a pressure of 10 bar when the condition of steam is*

1. *Wet and dryness fraction 0.85*
2. *Dry and saturated*
3. *Super heated, the degree of superheat being 50°C.*

The specific heat of superheated steam at constant pressure is 2.01 KJ/kg K. You can use steam table or Mollier's chart.

Solution. Given data

$$P = 10 \text{ bar}$$

By steam table

(1) Wet steam $x = 0.85$

From dry and saturated steam table corresponding to a pressure of 10 bar we get

$$u_f = 761.7 \text{ KJ/kg}$$
$$u_g = 2583.6 \text{ KJ/kg}$$
$$u_{fg} = u_g - u_f$$
$$= 2583.6 - 761.7$$
$$= 1821.9 \text{ KJ/kg}$$
$$u = u_f + x\,u_{fg}$$
$$= 761.7 + 0.85 \times 1821.9$$
$$= 2310.3 \text{ KJ/kg.}$$

(2) Dry and saturated steam

$$u_g = 2583.6 \text{ KJ/kg}$$

(3) Superheated steam

Degree of superheat = 50°C

From steam table, saturation temperature of steam at 10 bar is

$$t_{sat} = 180°\text{C}$$

Temperature of superheated steam

$$t_{sup} = t_{sat} + 50 = 180 + 50$$
$$= 230°\text{C.}$$

From superheated steam table corresponding to a pressure of 10 bar 230°C temperature,

$$u_g = 2674.7 \text{ KJ/kg.} \quad \textbf{Ans.}$$

Example 11.7. *1 kg of steam at 120 bar and 380°C expands isentropically to a pressure of 38 bar. Calculate the workdone by steam using (i) steam table (ii) Mollier chart.*

Solution. Process 1–2. Isentropic expansion ($s_1 = s_2$, $q = 0$)

$$m = 1 \text{ kg}$$
$$P_1 = 120 \text{ bar}$$
$$t_1 = 380°\text{C}$$
$$P_2 = 38 \text{ bar}$$

Solution.

(1). By using steam table

From steam table corresponding to 120 bar press, the saturation temperature of steam is

$$t_{sat} = 324.8°\text{C.}$$

Since the temperature is more than saturation temperature, it is superheated steam. From superheated steam table corresponding to a press of 120 bar and 380°C temperature.

$$s_1 = 5.963 \text{ KJ/kg°K}$$

$$h_1 = 2977.4 \text{ KJ/kg}$$

$$v_1 = 0.01965 \text{ m}^3\text{/kg}$$

$$u_1 = h_1 - P_1 v_1$$

$$= 2977.4 - 120 \times 100 \times 0.01965$$

$$= 2741.6 \text{ KJ/kg.}$$

From dry and saturated steam table corresponding to a pressure of 38 bar, we get,

$$s_{f2} = 2.769 \text{ KJ/kg K}$$

$$s_{fg2} = 3.323 \text{ KJ/kg K}$$

$$u_{f2} = 1068 \text{ KJ/kg}$$

$$u_{g2} = 2603 \text{ KJ/kg}$$

$$u_{fg2} = u_{g2} - u_{f2}$$

$$= 2603 - 1068$$

$$= 1535 \text{ KJ/kg}$$

For isentropic process $s_1 = s_2 = 5.963$ KJ/kg K

$$s_2 = s_{f2} + x_2 s_{fg2}$$

$$5.963 = 2.769 + x_2 \times 3.323$$

or $x_2 = 0.96$

$$u_2 = u_{f2} + x_2 u_{fg2}$$

$$= 1068 + 0.96 \times 1535$$

$$= 2541.6 \text{ KJ/kg.}$$

From first law of thermodynamics

$$q = w + u_2 - u_1$$

$$o = w + u_2 - u_1$$

or $w = u_1 - u_2 = 2741.6 - 2541.6$

$= 200$ KJ/kg **Ans.**

(2) By Mollier chart

Locate point 1 at the intersection of 120 bar press line and 380°C temperature line. Read the value of h_1 and v_1

$$h_1 = 2980 \text{ KJ/kg}$$

$$v_1 = 0.02 \text{ m}^3\text{/kg}$$

From point 1, draw a vertical line and mark point 2 where this line cuts the constant pressure line of 38 bar. Read the value of h_2 and v_2

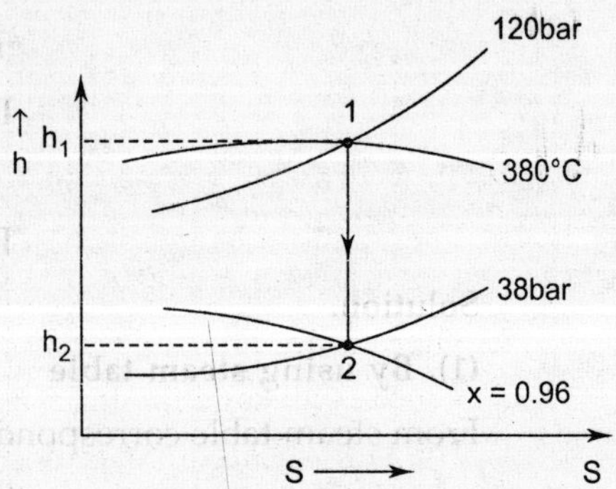

Fig. 11.18

$$h_2 = 2730 \text{ KJ/kg}$$
$$v_2 = 0.05 \text{ m}^3/\text{kg}$$

Internal energy is given by

$$u_1 = h_1 - P_1 v_1$$
$$= 2980 - 120 \times 100 \times 0.02$$
$$= 2740 \text{ KJ/kg}$$
$$u_2 = h_2 - P_2 v_2$$
$$= 2730 - 38 \times 100 \times 0.05$$
$$= 2540 \text{ KJ/kg}$$

workdone $w = u_1 - u_2$

$$= 2740 - 2540 = 200 \text{ KJ/kg} \quad \textbf{Ans.}$$

Example 11.8. *10 kg/s of superheated steam at 5 bar and 300°C expands through a turbine to final pressure of 0.1 bar and dryness fraction 0.95. find out the work generated by the turbine. Is the flow reversible? Explain.*

Solution. Given data

$$m = 10 \text{ kg/s}$$
$$P_1 = 5 \text{ bar}$$
$$t_1 = 300°\text{C}$$
$$P_2 = 0.1 \text{ bar}$$
$$x_2 = 0.95$$

From superheated steam table corresponding to a pressure of 5 bar and temperature 300°C,

$$h_1 = 3064.2 \text{ KJ/kg}$$

From dry and saturated steam table corresponding to press 0.1 bar. $h_{f2} = 191.8$ KJ/kg

$$h_{fg2} = 2392.8 \text{ KJ/kg}$$
$$h_2 = h_{f2} + x_2 h_{fg2}$$
$$= 191.8 + 0.95 \times 2392.8$$
$$= 2465 \text{ KJ/kg}$$

Work generated (power) by the turbine

$$w = m(h_1 - h_2)$$
$$= 10\,(3064.2 - 2465)$$
$$= 5992 \text{ kW.} \quad \textbf{Ans.}$$

The flow through the turbine is not reversible due to the following reasons.

1. There is loss of heat due to friction of steam with blades and casting of the turbine. The entropy of steam increases due to friction.
2. There is loss of heat due to radiations.
3. There is leakage of steam to atmosphere.
4. The superheated steam becomes wet at the end of expansion.

Example 11.9. *Steam initially at 1 bar pressure and 0.965 dryness fraction is throttled by passing through a constriction. Calculate (a) Dryness fraction at 0.5 bar pressure. (b) The pressure at which the steam would have become dry and saturated.*

Solution. Given data $P_1 = 1$ bar

$x_1 = 0.965$

$P_2 = 0.5$ bar

(*a*) Throttling process ($h_1 = h_2$)

From dry and saturated steam table corresponding to 1 bar pressure.

$$h_{f1} = 417.5 \text{ KJ/kg}$$

$$h_{fg1} = 2258 \text{ KJ/kg}$$

$$h_1 = h_{f1} + x_1 h_{fg1}$$

$$= 417.5 + 0.965 \times 2258$$

$$= 2596.4 \text{ KJ/kg}$$

From dry and saturated steam table corresponding to 0.5 bar pressure

$$h_{f2} = 340.6 \text{ KJ/kg}$$

$$h_{fg2} = 2305.4 \text{ KJ/kg}$$

$$h_2 = h_1 = 2596.47 = h_{f2} + x_2 h_{fg2}$$

$$= 340.6 + x_2 \times 2305.4$$

or $x_2 = 0.9785$. **Ans.**

(*b*) From dry and saturated steam table, the pressure at which the steam has enthalpy $h_g = 2596.47$ KJ/kg is 0.14 bar

$$P_3 = 0.14 \text{ bar.}$$

By Mollier chart

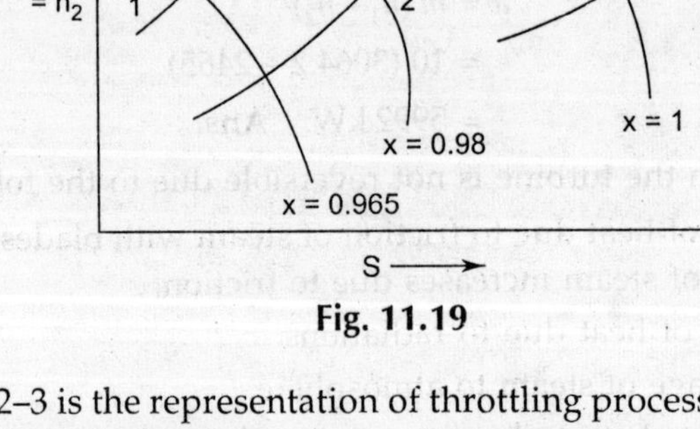

Fig. 11.19

The straight line 1–2–3 is the representation of throttling process, point 1 being the initial state of steam which is obtained by intersection of constant pressure line of 1 bar and constant dryness fraction line 0.965.

(a) Draw a horizontal line from point 1 and obtain point 2 at the intersection of this line with constant pressure line of 0.5 bar. Read the value of dryness fraction at this point. It is $x = 0.98$.

(b) Extend the horizontal line upto dry and saturated line ($x = 1$) to obtain point 3 at the intersection. Read the value of pressure at this point.

$$P_3 = 0.14 \text{ bar.} \quad \textbf{Ans.}$$

Example 11.10. *A steam main closed at both ends contains steam at a pressure of 8 bar and temperature of 200°C. It has an internal volume of 0.652 m³. If the press in the steam main falls to 3 bar due to heat loss from the steam pipe, find the mass of steam in the pipe and then quality and temperature of the steam at the final state.*

Solution. Given data

$$V = 0.652 \text{ m}^3$$

$$P_1 = 8 \text{ bar}$$

$$t_1 = 200°\text{C}$$

$$P_2 = 3 \text{ bar}$$

Process 1 – 2 constant volume ($v_1 = v_2$)

From dry and saturated steam table corresponding to 8 bar the saturation temperature of steam is 170.4°C.

$$\therefore \quad t_{sat} = 170.4°\text{C}$$

Since the temperature 200°C is more than 170.4°C, the steam is superheated.

From superheated steam table corresponding to 8 bar pressure and 200°C.

$$v_1 = 0.2608 \text{ m}^3/\text{kg}$$

$$\text{Mass of steam} = \frac{V}{v_1} = \frac{0.652}{0.2608} = 2.5 \text{ kg.} \quad \textbf{Ans.}$$

From dry and saturated steam table corresponding to 3 bar, $t_{sat} = 133.5°\text{C}$

$$v_{f2} = 0.001073 \text{ m}^3/\text{kg}$$

$$v_{g2} = 0.606 \text{ m}^3/\text{kg}$$

$$v_{fg2} = v_{g2} - v_{f2}$$

$$= 0.606 - 0.001073 = 0.604927 \text{ m}^3/\text{kg}$$

$$= 0.604927 \text{ m}^3/\text{kg}$$

$$v_2 = v_1 = v_{f2} + x_2\, v_{fg2}$$

or

$$0.2608 = 0.001073 + x_2 \times 0.604927$$

$$x_2 = 0.429$$

Temperature of steam at the final state = 133.5°C. **Ans.**

Quality of steam $x_2 = 0.429$. **Ans.**

Example 11.11. *A steam power plant works between pressure 30 bar and 0.04 bar. If the steam supplied is dry saturated and the cycle of operation is Rankine, calculate (i) cycle efficiency (ii) specific steam consumption.*

Solution. $\eta_{cycle} = 1 - \frac{h_4 - h_1}{h_3 - h_2}$

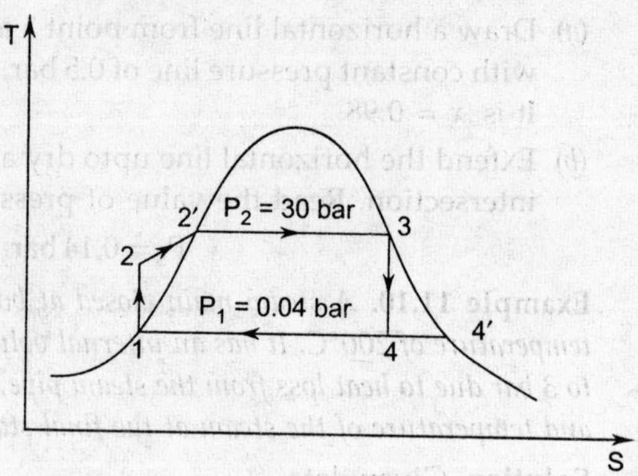

Fig. 11.20

From steam table II

At 30 bar

$$h_3 = h_{g3} = 2804.2 \text{ KJ/kg}$$

$$s_3 = s_{g3} = 6.187 \text{ KJ/kg}$$

At 0.04 bar

$$v_f = 0.001004 \text{ m}^3/\text{kg}$$

$$v_g = 34.80 \text{ m}^3/\text{kg}$$

$$h_f = 121.5 \text{ KJ/kg}$$

$$h_{fg} = 2432.9 \text{ KJ/kg}$$

$$h_g = 2554.4 \text{ KJ/kg}$$

$$s_f = 0.423 \text{ KJ/kg K}$$

$$s_{fg} = 8.052 \text{ KJ/kg K}$$

$$s_g = 8.475 \text{ KJ/kg K}$$

$$s_3 = s_4 = 6.187 = s_{f4} + x_4\, s_{fg4} = 0.423 + x_4\, 8.052$$

or $x_4 = 0.7158$ = Quality of wet steam at point 4.

$\therefore$ $h_4 = h_{f4} + x_4\, h_{fg4} = 121.5 + 0.7158 \times 2432.9 = 1863 \text{ KJ/kg}$.

$$h_1 = 121.5 \text{ KJ/kg}$$

w_p (pump work) $= v_{f1}(P_2 - P_1)$

$= 0.001004\,(30 - 0.04) \times 10^5 \text{ J/kg}$

$= 3.008 \text{ KJ/kg}$.

Again $w_p = h_2 - h_1$

or $3.008 = h_2 - 121.5$

or $h_2 = 124.508 \text{ KJ/kg}$.

(*i*) **Cycle efficiency :**

$$\eta_{cycle} = 1 - \frac{h_4 - h_1}{h_3 - h_2}$$

$$= 1 - \frac{1863 - 121.5}{2804.2 - 124.508} = 0.3501$$

= 35.01% **Ans.**

Turbine work $= w_T = h_3 - h_4 = 2804.2 - 1863$

$= 941.2 \text{ KJ/kg}$.

(*ii*) **Specific steam consumption:**

$$= \frac{3600}{\text{Turbine work}} = \frac{3600}{941.2} \text{ kg/kWh}$$

$= 3.824$ kg/kWhr.

Chapter **12**

Vapour Compression Refrigeration Cycle

12.1 INTRODUCTION

Vapour compression refrigeration system is the commonly used form of refrigeration system. It has commercial and domestic utility. The working fluid is vapour. It readily evaporates and condenses changing its phase from vapour to liquid and vice-versa. It absorbes heat from the body to be cooled during evaporation. Latent heat is used to convert liquid to vapour. During condensation it rejects heat to external body.

12.2 SIMPLE VAPOUR COMPRESSION SYSTEM

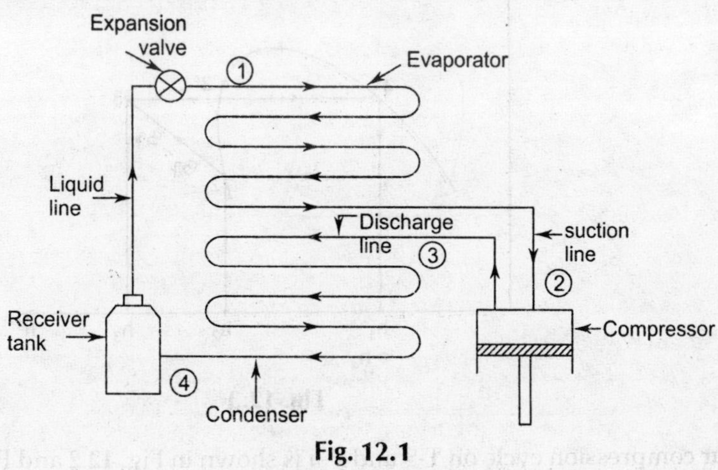

Fig. 12.1

Fig. 12.1 shows a simple vapour compression refrigeration system which consists of
• compression • condensation • Expansion • Vaporisation.

(*a*) **Compression :** The vapour at low pressure and low temperature (point 2) enters to compressor where it is compressed reversible adiabatically (isentropically). The pressure and temperature increase and pass through discharge line (point 3).

(*b*) **Condensation :** The vapour at high pressure and temperature enters to condenser where it is condensed to high pressure liquid (point 4) and is collected by receiver tank.

(*c*) **Expansion :** The liquid from receiver tank passes through the expansion valve where it is throttled (isoenthalpy) and passed at a controlled rate after reducing its pressure and temperature.

(*d*) **Vaporisation :** In evaporator the liquid at low pressure and temperature evaporates and is changed into vapour. During vaporisation the liquid absorbs latent heat of vaporisation from the medium (air, water or brine) which is to be cooled.

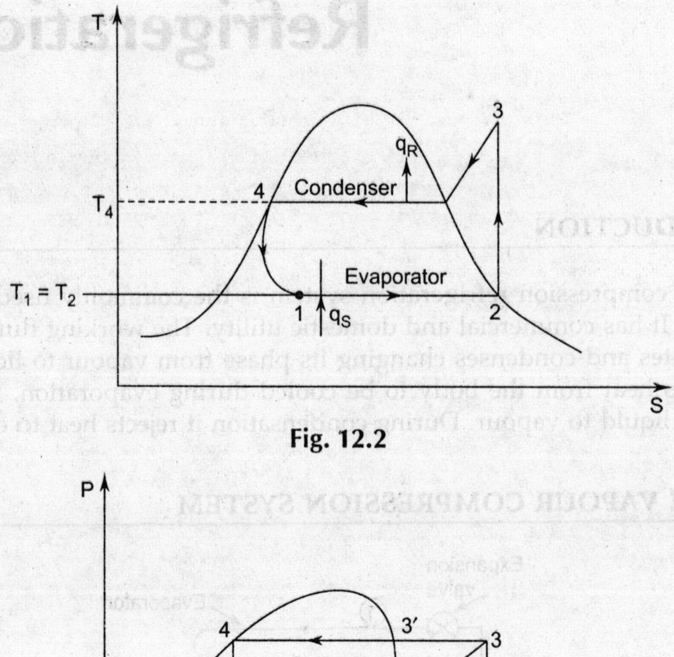

Fig. 12.2

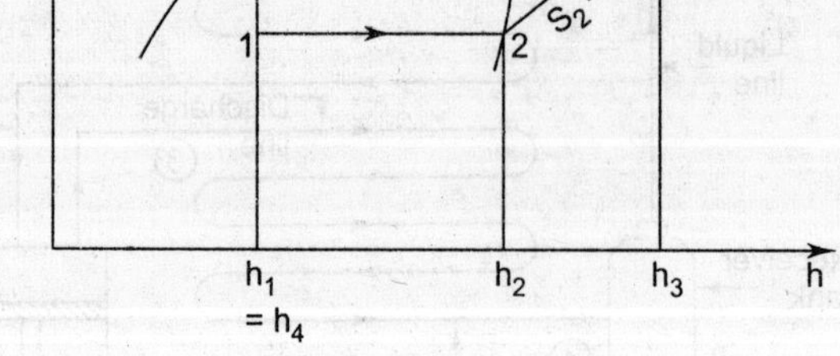

Fig. 12.3

Vapour compression cycle on T-S and P-*h* is shown in Fig. 12.2 and Fig. 12.3 respectively.

12.2.1 Coefficient of Performance of Refrigerator for Vapour Compression Cycle

Coefficient of Performance of refrigerator for vapour compression cycle is given by

$$\text{COP} = \frac{h_2 - h_1}{h_3 - h_2}, \left(\text{COP} = \frac{\text{heat rejected by evaporator}}{\text{work input to compressor}} \right)$$

12.2.2 The Quantity of Refrigerant (m_r) Circulated in the Refrigeration Plant

The Quantity of refrigerant is given by

$$m_r = \frac{210\,\text{T}}{h_2 - h_1}\ \text{kg/min}$$

where T = Load in tons of refrigeration on the refrigerator

12.2.3 Power Required to Drive the Compressor

$$\text{Power} = \frac{m_r\,(h_3 - h_2)}{60}\ \text{KW}$$

where h is in KJ/kg.

12.2.4 Quantity of Cooling Water (m_c) Circulated in the Condenser

Quantity of cooling water circulated the condenser per minute can be found out from the equation $m_c\ C_c\ (t_o - t_i) = m_r\ (h_3 - h_4)$.

where m_c = mass of cooling water circulated per minute

C_c = specific heat of cooling water

t_o = outlet temperature of cooling water

t_1 = inlet temperature of cooling water.

12.3 PRESENTATION OF PROCESSES ON P–h DIAGRAM

Fig. 12.4 shows the various processes on P-h diagram.

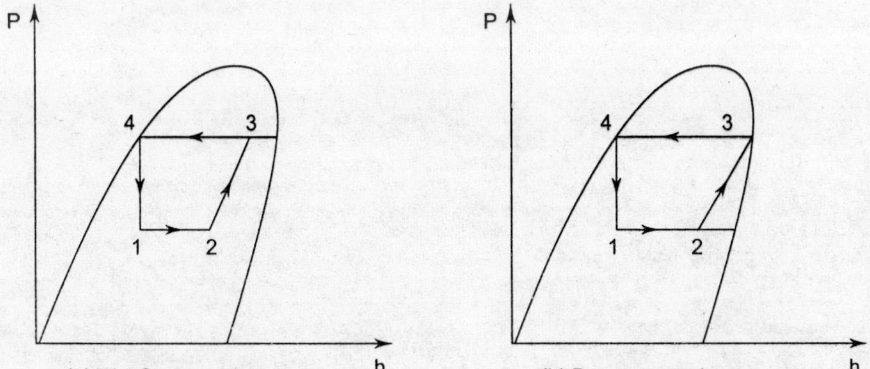

(a) Wet Compression (Vapour is wet at suction)

(b) Dry compression (Vapour is wet at suction)

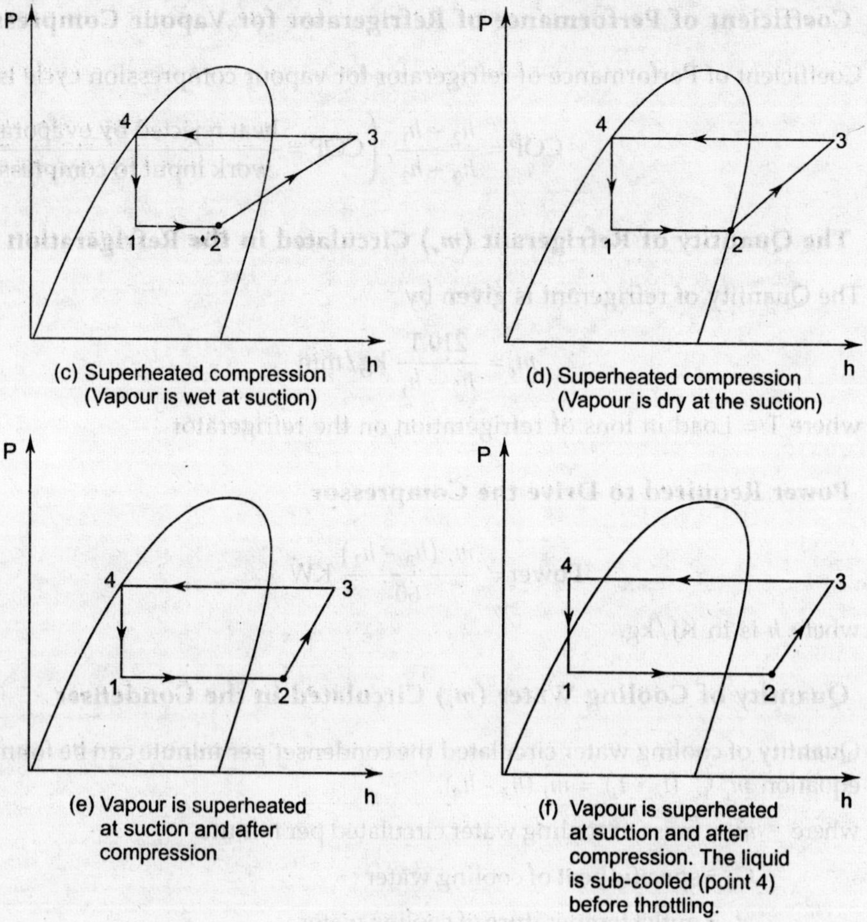

(c) Superheated compression (Vapour is wet at suction)

(d) Superheated compression (Vapour is dry at the suction)

(e) Vapour is superheated at suction and after compression

(f) Vapour is superheated at suction and after compression. The liquid is sub-cooled (point 4) before throttling.

Fig. 12.4

SECTION B

BASIC FLUID MECHANICS

Chapter 1

Properties of Fluids

1.1 INTRODUCTION

Fluid Mechanics is that branch of engineering science which deals with static, Kinematics and dynamic behaviour of fluids and fluid systems. When the fluid is at rest, the study of it is called fluid statics and when the fluid is in motion, the study of it without considering forces, is called Kinematics. If we consider forces causing the motion, the study of the fluid is termed as fluid dynamics. Liquid and gases are in the family of fluid.

1.2 DEFINITION OF FLUID

A substance that deforms continuously under the action of shear stress is called a fluid.

1.3 PROPERTIES OF FLUIDS

The physical properties of fluids are as follows:

1.3.1 Mass Density or Density

Mass density or simply density of a fluid is defined as the mass per unit volume. It is denoted as ρ (rho). If Δm and $\Delta\forall$ are mass and volume respectively of a fluid element enclosing a point P, then ρ is mathematically written as

$$\rho = \lim_{\Delta\forall \to 0} \frac{\Delta m}{\Delta\forall} = \left.\frac{dm}{d\forall}\right|_P$$

Fig. 1.1 shows a fluid element enclosing point P.

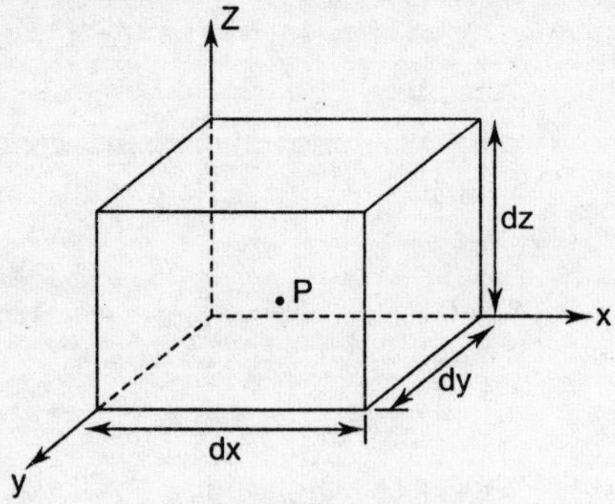

Fig. 1.1

$$dm = \rho\, d\forall = \rho\, dx\, dy\, dz.$$

S.I. (Standard International) unit of ρ is given by

$$\rho = \frac{dm}{d\forall} = \frac{kg}{m^3}.$$

$$\rho_{water} = 1000 \text{ kg/m}^3.$$

1.3.2 Specific Weight or Weight Density

The specific weight or weight density of a fluid is defined as the weight of the fluid per unit volume. It is denoted as w. It is mathematically expressed as

$$w = \left.\frac{dW}{d\forall}\right|_P = \left.\frac{g\,dm}{d\forall}\right|_P = g\left[\frac{dm}{d\forall}\right]_P$$

or $\quad w = g\rho$ or $\boxed{w = \rho g}$

where $\quad dW$ = weight of fluid element.

$= gdm.$

$$w_{water} = 9.81 \times 1000 \text{ N/m}^3$$

In S.I. unit $w = \dfrac{dW}{d\forall} = \text{N/m}^3$ [dW in N(newton), $d\forall$ in m^3.]

1.3.3 Specific Volume

The volume of a fluid per unit mass is called specific volume. It is denoted as v. It is mathematically expressed as

$$v = \left.\frac{d\forall}{dm}\right|_P = \frac{1}{\left.\dfrac{dm}{d\forall}\right|_P} = \frac{1}{\rho}$$

$$\boxed{v = \frac{1}{\rho}}$$

In S.I unit it is m^3/kg. We encounter v frequently in case of gases only.

1.3.4 Specific Gravity

The specific gravity of a liquid is defined as the ratio of weight density or density of the liquid to the weight density or density of pure water at pressure 101 KN/m² and temperature at 4°C. It is denoted as S. It is mathematically expressed as

$$S = \frac{w_{\text{liquid}}}{w_{\text{std.water}}} = \frac{w_{\text{liquid}}}{9.81 \times 1000\ N/m^3}$$

or $$w_{\text{liquid}} = S \times 9.81 \times 1000\ \text{N/m}^3.$$

S can also be expressed as

$$S = \frac{\rho_{\text{liquid}}}{\rho_{\text{std.water}}} = \frac{\rho_{\text{liquid}}}{1000\ kg/m^3}$$

or $$\rho_{\text{liquid}} = S \times 1000\ \text{kg/m}^3.$$

The S.I. unit of S: From its definition, we get S as dimensionless quantity.

1.4 CLASSIFICATION OF FLUIDS

1.4.1 Ideal and Real Fluids

Ideal fluid is defined as the fluid which is incompressible (ρ = constant) and having no viscosity (μ = 0). Ideal fluid is imaginary (because μ ≠ 0 in any fluid in real word) Real fluid is defined as the fluid which is having viscosity (μ ≠ 0) and compressible ($k \neq 0$).

1.4.2 Newtonian's Fluids and Newton's Law of Viscosity

The property of a fluid which by virtue of it, offers resistance to the motion of layers of fluid is called viscosity. When fluid flows over a solid surface, the velocity is different at different points along the depth. It is zero (no slip) at the solid surface and increases along depth. Fig. 1.2 shows when two layers of a fluid at a distance dy apart, move one over the other at different velocities u and $u + du$. Layer 2 offers a resistance to layer 1 and vice-versa. The upper layer 2 moving with velocity $(u + du)$ drags the lower layer 1 with velocity u causing a resistance of magnitude F Newton. The shear stress is developed which is F/A where A is the contact area between the two layers. Newton's law of viscosity states that shear stress is proportional to the spatial rate of change of velocity normal to the flow.

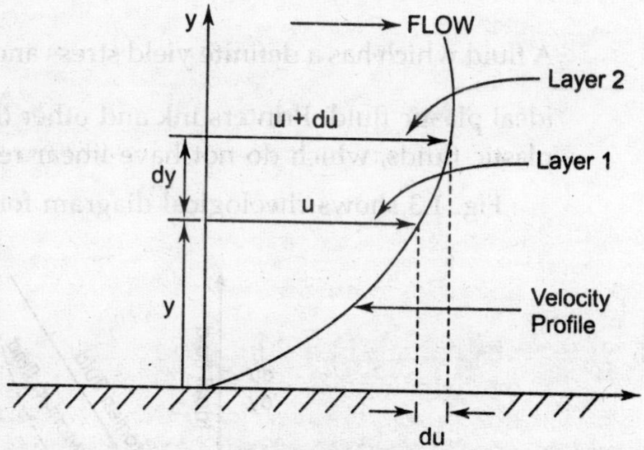

Fig. 1.2

Thus mathematically

$$\tau \propto \frac{du}{dy} \text{ or } \tau = \mu \frac{du}{dy} \quad ...(1.1)$$

where τ is the shear stress in N/m^2, du in m/s, dy in m. So the S.I. unit of μ (called mu) (coefficient of dynamic viscosity) is given by

$$\mu = \frac{\tau}{\frac{du}{dy}} = \frac{\frac{N}{m^2}}{\frac{m}{s} \times \frac{1}{m}} = \frac{N.s}{m^2}$$

The unit of viscosity (in C.G.S) is poise.

$$\text{One poise} = 0.1 \frac{N.s}{m^2}, \ \mu_{water} = 10^{-3} \frac{Ns}{m^2}$$

$$\mu_{air} = 1.81 \times 10^{-5} \ Ns/m^2.$$

Newtonian's Fluid

The fluid which obeys the law $\tau = \mu \frac{du}{dy}$ is called Newtonian's fluid. It is a real fluid. Water, air, mercury are Newtonian's fluids.

Non-Newtonian's Fluid

The fluid which does not obey the law $\tau = \mu \frac{du}{dy}$ is called Non-Newtonian's fluid. Paints, polymers, blood are Non-Newtonian's fluids.

1.4.3 Ideal Plastic Fluid

A fluid which has a definite yield stress and a constant linear relation of τ to $\frac{du}{dy}$ is called ideal plastic fluid. Printers ink and other thixotropic substance are the examples of real plastic fluids, which do not have linear relation of τ to du/dy.

Fig. 1.3 shows rheological diagram for different kinds of fluids.

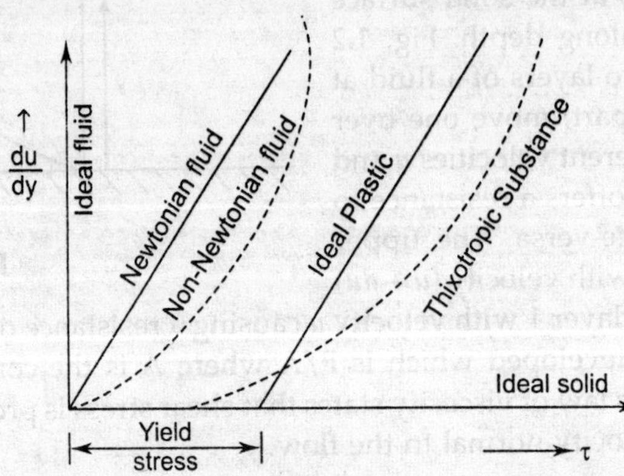

Fig. 1.3

1.5 KINEMATIC VISCOSITY

Kinematic viscosity is the dynamic viscosity per unit mass density. It is denoted as ν (called nu). It is mathematically expressed as

$$\nu = \frac{\mu}{\rho}$$

The S.I. unit of ν is given by

$$\nu = \frac{\frac{N.s}{m^2}}{\frac{kg}{m^3}} = \frac{Ns}{m^2} \times \frac{m^3}{kg} = \frac{kg\,m\,s}{s^2\,m^2}\frac{m^3}{kg}$$

$$= \frac{m^2}{s}.$$

[1 Stoke = 1 $cm^2/s = 10^{-4}\ m^2/s$.]

1.6 VARIATION OF DYNAMIC VISCOSITY WITH TEMPERATURE

For all fluids viscosity changes with temperature.

For liquids: $$\frac{\mu_0}{\mu} = 1 + \alpha t + \beta t^2 \qquad ...(1.2)$$

where μ_o = viscosity of liquid at 0°C, in poise
μ = viscosity of liquid at t°C in poise
α, β = constants for the liquid.

From equation (1.2) we see that viscosity of liquid decreases with increase of temperature.

For water: $\mu_o = 1.79 \times 10^{-3}$ poise, $\alpha = 33.7 \times 10^{-3}$, $\beta = 22.1 \times 10^{-5}$

For gases: $$\mu = \mu_o (1 + \alpha t - \beta t^2) \qquad ...(1.3)$$

For air: $\mu_o = 17 \times 10^{-6}$ poise, $\alpha = 56 \times 10^{-9}$, $\beta = 119 \times 10^{-12}$

From equation (1.3), we see that viscosity of gases increases with the increase of temperature because in case of gases the cohesive force is small and molecular momentum transfer predominates whereas in liquid molecular momentum is small and cohesive force predominates.

1.7 COMPRESSIBLE FLUIDS

A compressible fluid is characterised by its compressibility denoted as $\left(\frac{1}{K}\right)$, where K is the bulk modulus of elasticity.

Let

$\forall$ = initial volume of the gas

$\forall - d\forall$ = final volume of the gas after compression

change in volume = final volume − initial volume

$= (\forall - d\forall) - (\forall)$

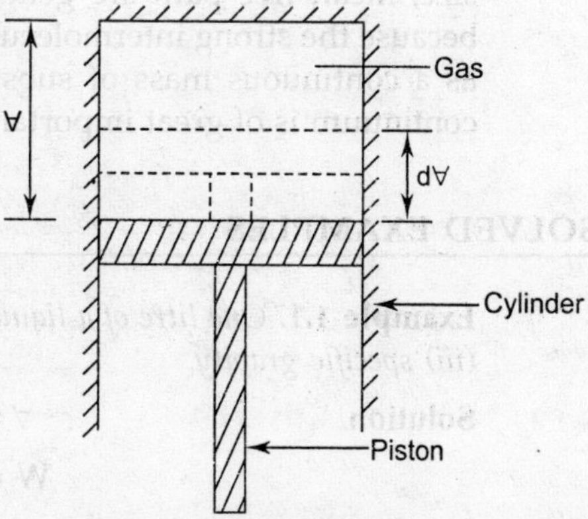

Fig. 1.4

$$= -d\forall$$

$$\text{volumetric strain} = \frac{\text{change in volume}}{\text{initial volume}} = \frac{-d\forall}{\forall}.$$

$$\text{Bulk modulus of elasticity} = K = \frac{dp}{\frac{-d\forall}{V}} \quad \text{where } dp = \text{increase in pressure}$$

$$\text{compressibility} = \frac{1}{K} = \frac{1}{\frac{dp}{\frac{-d\forall}{\forall}}} = \frac{\frac{-d\forall}{\forall}}{dp} = \frac{\frac{d\rho}{\rho}}{dp} = \left(\frac{1}{\rho}\frac{d\rho}{dp}\right) \quad \text{...(1.4)}$$

Gases are compressible fluids. [**Note.** $m = \rho\forall$, or $\forall d\rho + \rho\, d\forall = 0$ or $\frac{d\rho}{\rho} = -\frac{d\forall}{\forall}$.]

1.8 INCOMPRESSIBLE FLUIDS

The fluid in which compressibility is zero, is called incompressible fluid. From equation (1.4) we get,

$$0 = \frac{1}{\rho}.\frac{d\rho}{dp} \text{ or } d\rho = 0 \text{ or } \rho = \text{constant.}$$

Thus for incompressible fluid the mass density is constant.

1.9 CONCEPT OF CONTINUUM

In fact all fluids consist of discrete molecules but we consider a continuous and homogeneous fluid medium where the distribution of mass within the matter or system without empty space, is continuous. A continuum is a continuous and homogeneous medium where fluid properties such as density, viscosity, thermal conductivity, temperature, etc., can be expressed as continuous function of space and time. molecular size, mean free path are generally insignificant in order to analyse the fluid motion because the strong intermolecular cohesive forces make the entire liquid mass to behave as a continuous mass of substance. So in the study of liquid motion the concept of continuum is of great importance.

SOLVED EXAMPLES

Example 1.1. *One litre of a liquid weighs 8 N. Calculate its (i) specific weight (ii) density and (iii) specific gravity.*

Solution. $\forall = 1 \text{ litre} = 10^{-3} \text{ m}^3.$

$W = 8$ N

(i) $w = \frac{W}{\forall} = \frac{8}{10^{-3}}$ N/m^3 $= 8 \times 10^3$ N/m^3 **Ans.**

(ii) $\rho = \frac{w}{g} = \frac{8 \times 10^3}{9.81} = 8.155 \times 10^2$ kg/m^3 **Ans.**

(iii) $S = \frac{\rho}{1000} = \frac{8.155 \times 10^2}{10^3} = 0.8155$ **Ans.**

Example 1.2. *AB and CD are horizontal plates. AB is fixed CD moves with a velocity of 0.6 m/s and requires a force of 4 N per square meter. The distance between the plates is 0.03 mm. Find out the dynamic viscosity of the liquid between the plates.*

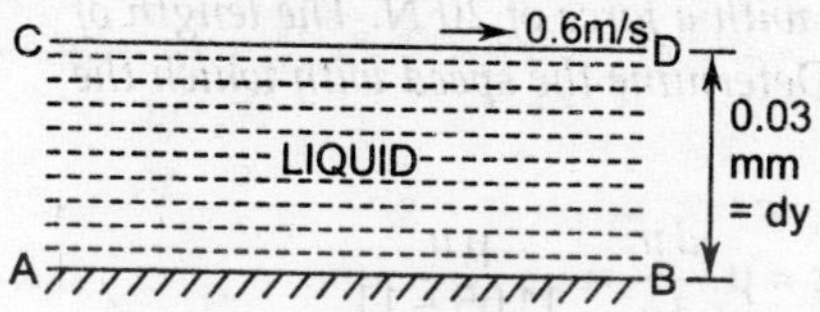

Fig. 1.5

Solution.

$$dy = 0.03 \text{ mm} = 0.03 \times 10^{-3} \text{ m}$$

$$du = 0.6 \text{ m/s} - 0 \text{ m/s} = 0.6 \text{ m/s}.$$

$$\tau = \frac{F}{A} = \frac{4\text{N}}{1m^2} = 4 \text{ N/m}^2$$

$$\tau = \mu.\frac{du}{dy}, \text{ or } 4 = \mu \times \frac{0.6}{0.03 \times 10^{-3}}$$

or

$$\mu = \frac{4 \times 0.03 \times 10^{-3}}{0.6} = 0.2 \times 10^{-3} \text{ Ns/m}^2$$

$$= 2 \times 10^{-3} \text{ poise. } \textbf{Ans.}$$

Example 1.3. *The upper plate is placed 0.2 mm apart from the lower plate. The space between them is filled with a viscous fluid of dynamic viscosity 2 poise. The upper plate of area 2 m^2 is required to move with a speed of 0.65 m/s relative to the lower plate. Determine the necessary (i) shear resistance or force and (ii) power required to maintain this speed.*

Solution.

$$dy = 0.2 \text{ mm}$$

$$= 0.2 \times 10^{-3} \text{ m}$$

$$A = 2 \text{ m}^2$$

$$\mu = 2 \text{ poise} = 0.2 \text{ Ns/m}^2$$

$$du = u_2 - u_1 = 0.65 \text{ m/s}$$

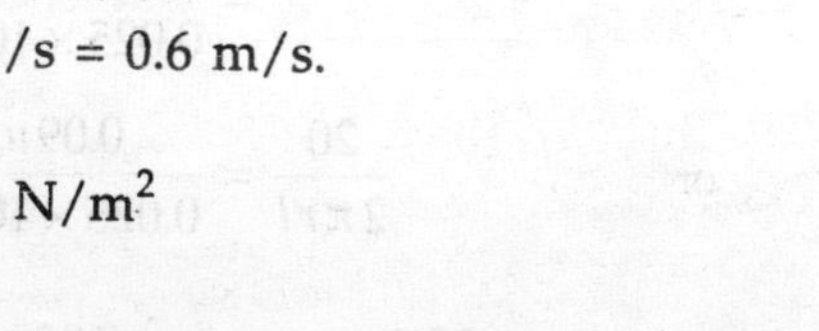

Fig. 1.6

$$\tau = \mu.\frac{du}{dy} = 0.2 \times \frac{0.65}{0.2 \times 10^{-3}}$$

(*i*) Shear resistance or force, $F = \tau A = \dfrac{0.2 \times 0.65 \times 2}{0.2 \times 10^{-3}}$

$= 1.3 \times 10^3$ N **Ans.**

(*ii*) Power, $P = F(u_2 - u_1) = 1.3 \times 10^3 (0.65)$ Nm/s

$= 845$ Nm/s

$= 845$ watt. **Ans.**

Example 1.4. *Fig. 1.7 shows oil of viscosity 0.9 poise filled annular space of a cylinder of 11.05 cm diameter. The piston's diameter is 11 cm. The piston moves vertically downward with a force of 20 N. The length of the piston is 13 cm. Determine the speed with which the piston slides down.*

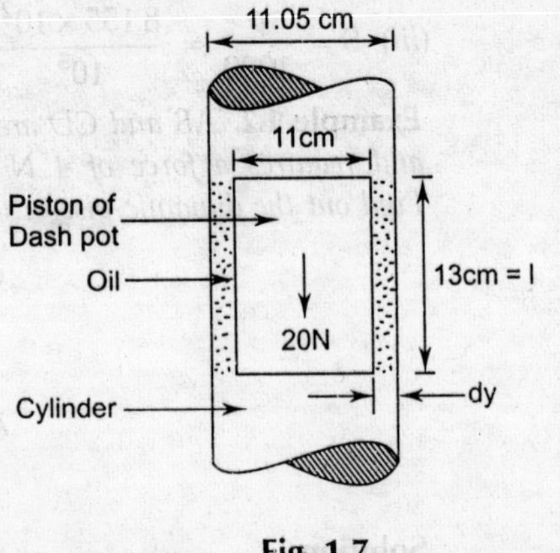

Fig. 1.7

Solution.

$$\tau = \mu.\frac{du}{dy} = \frac{\mu u}{\frac{11.05 - 11}{200}}$$

$$= \frac{0.09 \times u}{0.025 \times 10^{-2}}$$

or $$\frac{20}{2\pi r l} = \frac{0.09u}{0.025 \times 10^{-2}} \qquad [A = 2\pi r l]$$

or $$\frac{20}{2\pi \times \frac{11}{200} \times \frac{13}{100}} = \frac{0.09u}{0.025 \times 10^{-2}} \qquad \left[\begin{array}{l} r = \frac{11}{2} cm \\ l = 13 cm \end{array}\right]$$

$$u = \frac{20 \times 0.025 \times 10^{-2}}{2\pi \times \frac{11}{200} \times \frac{13}{100} \times 0.09} = 1.236 \text{ m/s.} \quad \textbf{Ans.}$$

Example 1.5. *A solid cube weighing 5 N and having a 45 Cm edge is allowed to slide down an inclined plane surface making an angle of 30° with the horizontal. There is a uniform oil film of 0.008 cm thick. If the cube is having a velocity of 13 cm/s, determine the viscosity of the oil. Also find out the Kinematic viscosity in stokes if the oil has a density of 850 kg/m³.*

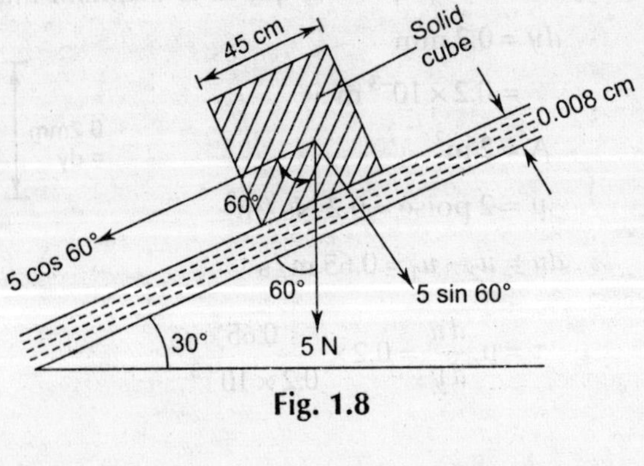

Fig. 1.8

Solution.

The tangential force along the inclined plane

$$= 5 \cos 60° = 2.5 \text{ N}$$

$$\tau = \mu.\frac{du}{dy}$$

$$\frac{2.5}{0.45 \times 0.45} = \mu \times \frac{0.13}{0.008 \times 10^{-2}},$$

$$\mu = 7.6 \times 10^{-3} \text{ Ns/m}^2$$

$$= 7.6 \times 10^{-2} \text{ poise} \quad \textbf{Ans.}$$

$$\nu = \frac{\mu}{\rho} = \frac{7.6 \times 10^{-3}}{850} = 0.894 \times 10^{-5} \text{ m}^2/\text{s}$$

$$= 8.94 \times 10^{-2} \text{ stokes. } \textbf{Ans.}$$

Example 1.6. *A journal bearing having 60 cm length has a clearance of 0.15 cm. The diameter of the shaft is 30 cm and it rotates at 4000 rpm. The viscosity of the newtonian lubricant used is 0.02 Ns/m² and the velocity variation in the lubricant is linear. Calculate the frictional torque the journal has to overcome, and the power loss.*

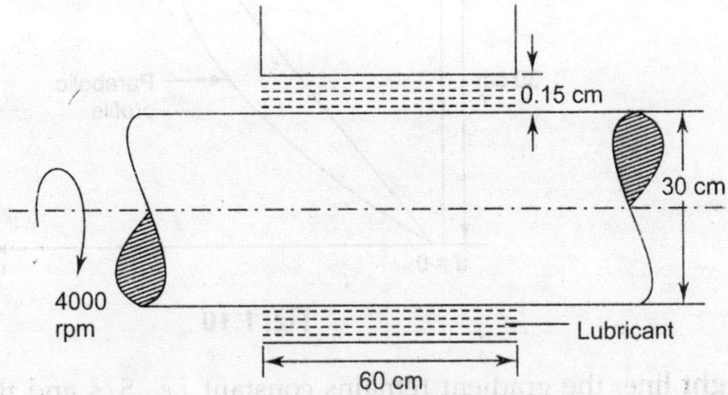

Fig. 1.9

Solution.

$$dy = 0.15 \text{ cm} = 0.15 \times 10^{-2} \text{ m}$$

$$\mu = 0.02 \text{ Ns/m}^2$$

$$u = \frac{2\pi N r}{60} = \frac{2\pi \times 4000 \times 0.15}{60} = 62.8 \text{ m/s} \text{ [Here } du = u - o = u]$$

$$\tau = \mu \times \frac{du}{dy} = 0.02 \times \frac{62.8}{0.15 \times 10^{-2}} = 837.3 \text{ N/m}^2$$

$$\text{force} = \tau A = 837.3 \times 2\pi r l$$

or $$F = 837.3 \times 2\pi \times 0.15 \times 0.60$$

or $$F = 473.48 \text{ N}.$$

$$\text{Torque due to friction} = F.r = 473.48 \times 0.15$$

$$= 71.02 \text{ Nm}$$

$$\text{Power loss} = F \times u = 473.48 \times 62.8 \text{ watt}$$

$$= 29.73 \text{ kW.} \quad \textbf{Ans.}$$

Example 1.7. *The velocity profile of a fluid of viscosity of 9 poise flowing over a fixed plate is assumed to be (i) straight line (ii) a parabola of vertex at 30 cm from the plate where the velocity is 150 cm/s. Calculate the velocity gradients and shear stress at points 4, 10 and 20 cm from the boundary. Write your observations on results of the two cases.*

Solution. Case (*i*)

$$\frac{du}{dy} = \frac{150 - 0}{30 - 0} = 5/\text{s}.$$

$$\tau = \mu.\frac{du}{dy} = 0.9 \times 5 \text{ Ns/m}^2 \times \frac{1}{s}$$

$$= 4.5 \text{ N/m}^2$$

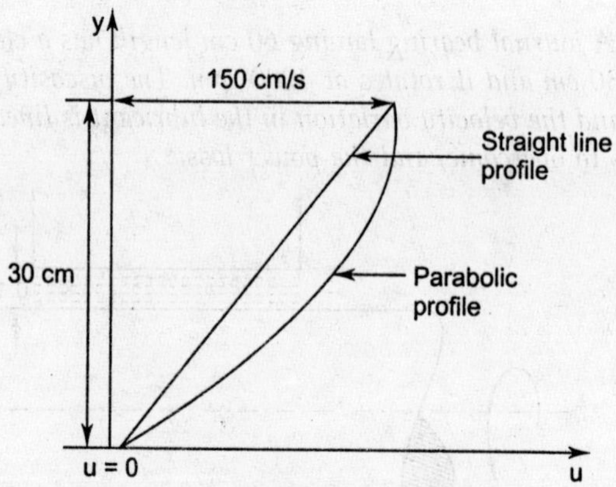

Fig. 1.10

For straight line, the gradient remains constant *i.e.*, 5/s and thus the shear stress also remains constant *i.e.* 4.5 N/m^2.

Case (*ii*) Let us assume the parabolic distribution as

$$u = ay^2 + by + c \qquad ...(1)$$

Thus $$\frac{du}{dy} = 2ay + b \qquad ...(2)$$

Boundary conditions at $y = 0$, $u = 0 \Rightarrow$ from (1), $c = 0$

at $y = 0.3$ m, $u = 1.5$ m/s

from (1) $\Rightarrow 1.5 = a\,(0.3)^2 + b\,(0.3) \Rightarrow a \times 0.3 + b = 5$ ...(3)

and $$\frac{du}{dy} = 0 \text{ at } y = 0.3 \text{ m}$$

from (2) $\Rightarrow 0 = 2a\,(0.3) + b$

or $$b = -0.6a \qquad ...(4)$$

from (3) $\Rightarrow b = 5 - 0.3a$

Thus $\quad -0.6\,a = 5 - 0.3a$

or $\quad 0.3a = -5$

or $\quad a = -16.7$

$$b = -0.6a = 16.7 \times 0.6 = 10.02.$$

Thus the parabolic profile is

$$u = -16.7y^2 + 10.02\,y$$

$$\frac{du}{dy} = 2ay + b = 2\,(-16.7)\,y + 10.02$$

$$= -33.4\,y + 10.02$$

for $\quad y = 4 \text{ cm} = 0.04 \text{ m}$

$$\frac{du}{dy} = -33.4\,(0.04) + 10.02$$

$$= 8.684 \text{ per sec.}$$

$$\tau_{y=0.04} = \mu.\frac{du}{dy} = 0.9 \times 8.684 = 7.815 \text{ N/m}^2$$

for $\quad y = 10 \text{ cm} = 0.1 \text{ m.}$

$$\frac{du}{dy} = -33.4\,(0.1) + 10.02 = 6.68 \text{ per sec}$$

$$\tau_{y=0.1\text{m}} = 6.68 \times 0.9 = 6.012 \text{ N/m}^2$$

for $\quad y = 20 \text{ cm} = 0.2 \text{ m}$

$$\frac{du}{dy} = -33.4\,(0.2) + 10.02$$

$$= 3.34 \text{ per sec.}$$

$$\tau_{y=0.2\text{m}} = 3.34 \times 0.9 = 3.006 \text{ N/m}^2$$

for $\quad y = 0,$

$$\frac{du}{dy} = 10.02 \text{ per sec.}$$

$$\tau_{y=0} = 10.02 \times 0.9 = 9.018 \text{ N/m}^2.$$

Observations:

(*i*) In parabolic distribution the velocity gradient $\left(\dfrac{du}{dy}\right)$ gets smaller and smaller values with increases in y. So we can say it is less steep.

(*ii*) The shear stress is having maximum value at the boundary and it decreases with increase in y.

Chapter 2

Fluid Statics

2.1 INTRODUCTION

The study of a fluid at rest is called fluid statics. As there is no relative motion between two layers of a fluid, the shear force is zero. In fluid statics we establish the relationship of pressure with the depth of the fluid.

2.2 FLUID PRESSURE AT A POINT

Let us consider a small area dA where the force dF is acting. Pressure is defined as the force per unit area.

So $$p = \frac{dF}{dA}.$$

If the force F is acting uniformly over the area A, then the pressure p is given by

$$p = \frac{F}{A}$$

The S.I. unit of pressure is given by

$$p \Rightarrow \frac{N}{m^2} \Rightarrow P_a \text{ (Pascal)}$$

Sometimes bar is used for pressure unit.

$$1 \text{ bar} = 10^5 \text{ N/m}^2 = 10^5 \text{ P}_a = 100 \text{ KP}_a.$$

and $$1 \text{ atm} = 101.3 \text{ KP}_a.$$

$$1 \text{ torr} = 1 \text{ mm of Hg}.$$

2.3 PASCAL'S LAW

Statement: *"In a static fluid the pressure or intensity of pressure at a point is equal in all directions".*

Proof: Let us take a fluid element as shown in Fig. 2.1. BC = dx, AB = dy, AE = dz AC = ds.

P_x = Pressure on ABOE surface

P_y = Pressure on BCDO surface

P_θ = Pressure on ACDE surface

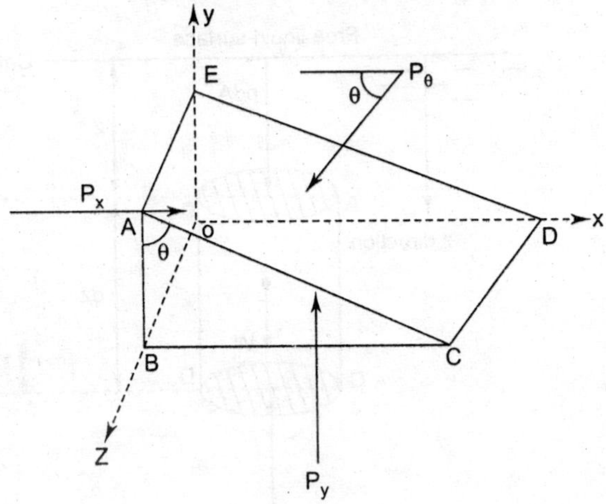

Fig. 2.1

∴ Force on ABOE = $P_x\, dy\, dz$

Force on BCDO = $P_y\, dx\, dz$

Force on ACDE = $P_\theta\, dz\, ds$.

As the fluid element is in equilibrium,

$$\Sigma F_x = 0 \qquad \text{[All forces on } x \text{ direction is zero]}$$

or $$P_x\, dy\, dz - P_\theta\, dz\, ds \cos\theta = 0$$

or $$P_x\, dy\, dz - P_\theta\, dz\, dy = 0 \qquad [ds \cos\theta = dy]$$

or $$P_x = P_\theta$$

And $$\Sigma F_y = 0 \qquad \text{[All forces on } y \text{ direction is zero]}$$

or $$P_y\, dx\, dz - P_\theta\, dz\, ds \sin\theta = 0$$

or $$P_y\, dx\, dz - P_\theta\, dz\, dx = 0 \qquad [ds \sin\theta = dx]$$

or $$P_y = P_\theta$$

Thus $$P_x = P_y = P_\theta \qquad \text{...(2.1)}$$

Equation (2.1) is independent of θ, therefore it is proved that pressure acts equally in all directions in a stationary fluid.

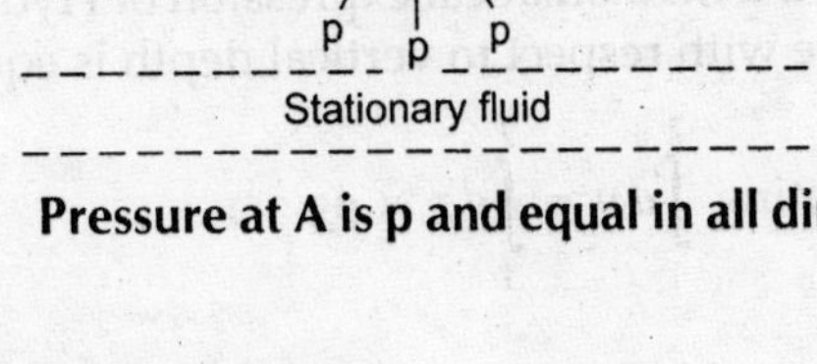

Fig. 2.2 Pressure at A is p and equal in all directions.

2.4 PRESSURE-DENSITY-HEIGHT RELATIONSHIP

Let us take an imaginary cylindrical fluid element which is at rest under the action of forces as shown in Fig 2.3.

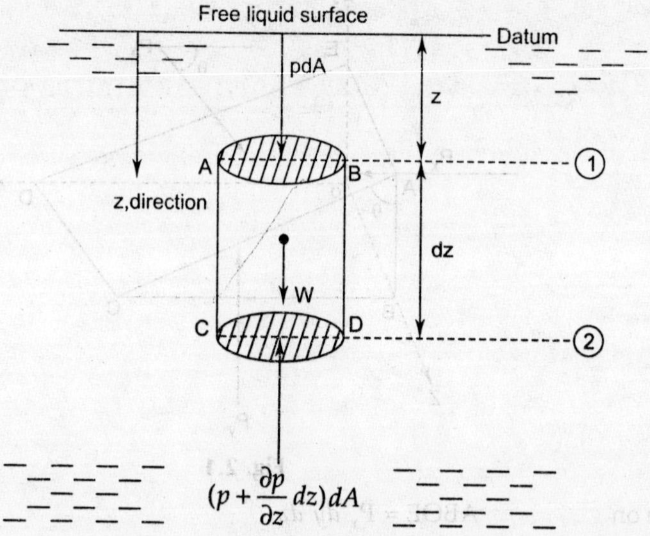

Fig. 2.3

$$\Sigma F_Z = 0$$

or $$PdA + W - \left(P + \frac{\partial p}{\partial z}dz\right)dA = 0$$

or $$\frac{\partial p}{\partial z}dz\,dA = W$$ [W is the weight of the fluid element]

or $$\frac{\partial p}{\partial z}dz\,dA = w\,dA\,dz,$$ [w = weight density]

or $$\frac{\partial p}{\partial z} = w$$

or $$\frac{dp}{dz} = w$$ [as p is a function of z only]

or $$\frac{dp}{dz} = \rho g \text{ [}\rho\text{ = mass density]} \quad ...(2.2)$$

Equation (2.2) is the mathematical expression of Hydrostatic law which states that rate of change of pressure with respect to vertical depth is equal to weight density of the fluid.

Integrating $$\int_1^2 dp = \int_1^2 dz \times \rho g$$

or $$p_2 - p_1 = \rho g (z_2 - z_1)$$

or $$\boxed{p = \rho g z}$$

where p is the pressure above atmospheric pressure and z is the height of the point from the free liquid surface or pressure head.

2.5 ATMOSPHERIC, GAUGE, VACUUM AND ABSOLUTE PRESSURE

Datum is at absolute zero line $\quad P_{abs}$ = Absolute pressure = $P_g + P_{at}$

and $\quad P_{abs} = P_{at} - P_{vac}$

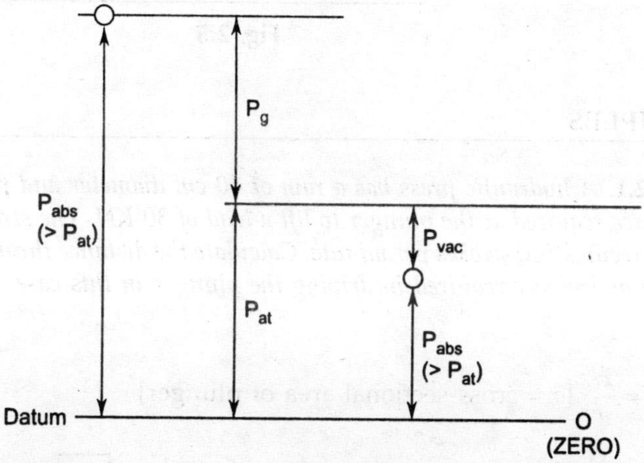

Fig. 2.4

P_{abs} = pressure measured from absolute zero pressure is called absolute pressure

P_g = Gauge pressure

= Pressure measured above the atmospheric pressure is called gauge pressure, here atmospheric pressure is taken as zero.

P_{vac} = Vacuum pressure = Pressure below the atmospheric pressure is the vacuum pressure.

P_{at} = Atmospheric Pressure

2.6 STATIC (P_s) AND TOTAL PRESSURE (P_t)

A fluid is flowing through a pipe. The static pressure is defined as the pressure on the wall by a static fluid. For fluid flow situation as shown in Fig 2.5, we can measure the static pressure by just putting a vertical tube in the wall of the pipe. If we like to measure the total pressure we have to fit a tube (pitot tube) as shown in Fig. 2.5 (right side tube). P_d is the indicated dynamic pressure which is the head difference between the two tubes.

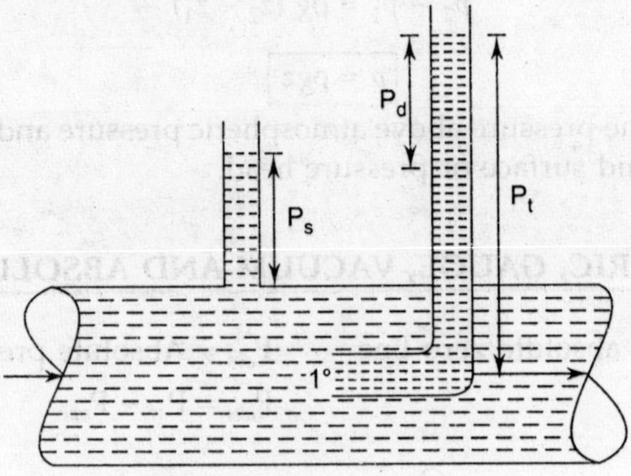

Fig. 2.5

SOLVED EXAMPLES

Example 2.1. *A hydraulic press has a ram of 40 cm diameter and plunger of 10 cm diameter. Find the force required at the plunger to lift a load of 30 KN. The stroke length of plunger is 40 cm and it executes 120 strokes per minute. Calculate the distance through which the load is raised per minute and power required in driving the plunger in this case.*

Solution.

(*a*) $p_1 = \frac{F}{a}$, [a = cross-sectional area of plunger]

$p_2 = \frac{W}{A}$, [A = cross-sectional area of ram]

$p_1 = p_2$

or $\frac{F}{a} = \frac{W}{A}$ or $W = F.\frac{A}{a}$

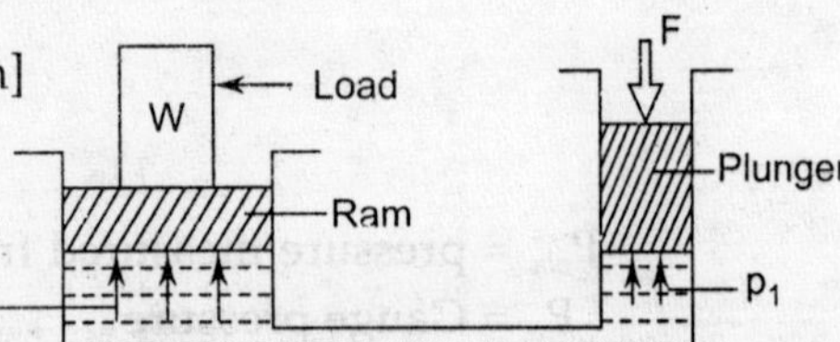

Fig. 2.6

or $F = W.\frac{a}{A} = 30 \times 10^3 \times \frac{\frac{\pi}{4}(0.1)^2}{\frac{\pi}{4}(0.40)^2} = 1875$ N. **Ans.**

(*b*) Stroke volume (plunger side) $= 0.4 \times \frac{\pi}{4}(0.1)^2$ m³.

Stroke volume (ram side) $= S_r \times \frac{\pi}{4}(0.4)^2$ m³.

$\therefore \quad S_r \times \frac{\pi}{4}(0.4)^2 = 0.4 \times \frac{\pi}{4}(0.1)^2$

or $\quad S_r = \frac{0.4\,(0.1)^2}{(0.4)^2} = 0.025$ m.

Distance raised on ram side

$= 0.025 \times 120$ m/minute = 3 m/minute **Ans.**

(*c*) Power required $= 30 \times 10^3 \times 3$ Nm/minute

$$= 9 \times 10^4 \text{ Nm/minute} = \frac{9}{60} \times 10^4 \text{ watt}$$
$$= 1.5 \text{ KW} \quad \textbf{Ans.}$$

Example 2.2. *In Fig. 2.7 find the pressure at A and also at B.*

Solution.

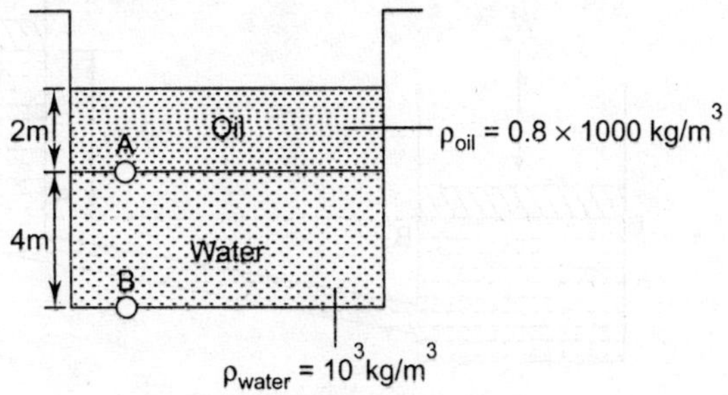

Fig. 2.7

Pressure at A, $\quad p_A = \rho_{oil} \times g \times z_1, \qquad (z_1 = 2\text{m})$

$$= 0.8 \times 1000 \times 9.81 \times 2$$
$$= 15696 \text{ N/m}^2 \quad \textbf{Ans.}$$

Pressure at B $= p_A + \rho_{water} \times g \times z_2, \qquad (z_2 = 4\text{m})$

$$= 15696 + 1000 \times 9.81 \times 4$$
$$= 54936 \text{ N/m}^2 \quad \textbf{Ans.}$$

Example 2.3. *The diameters of a small piston and a large piston are 2 cm and 8 cm respectively. A force of 100 N is applied on the small piston of the hydraulic jack. Calculate the load lifted by the large piston when (i) the pistons are at same level (Fig. 2.8) (ii) small piston is 50 cm above the large piston (Fig. 2.9).*

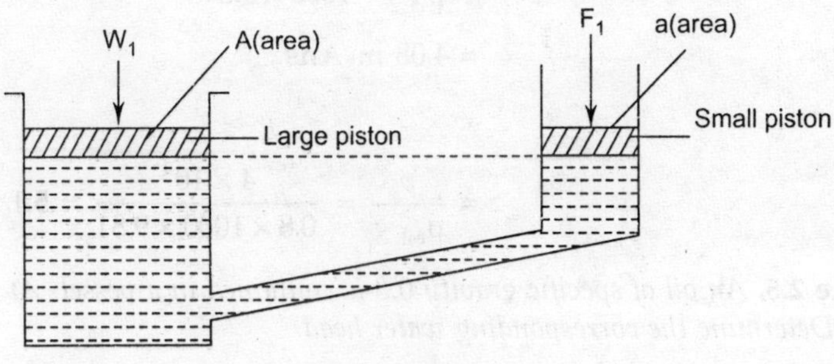

Fig. 2.8

Solution. (*i*)

$$W_1 = F_1.\frac{A}{a} = \frac{100 \times \frac{\pi}{4}(0.08)^2}{\frac{\pi}{4}(0.02)^2} = 1600 \text{ N} \;\; \textbf{Ans.}$$

(*ii*)

$$p_{AA} = 0.5 \times 1000 \times 9.81 + \frac{100}{\frac{\pi}{4}(0.02)^2}$$

$$= (4905 + 318310) \text{ N/m}^2$$
$$= 323215 \text{ N/m}^2$$

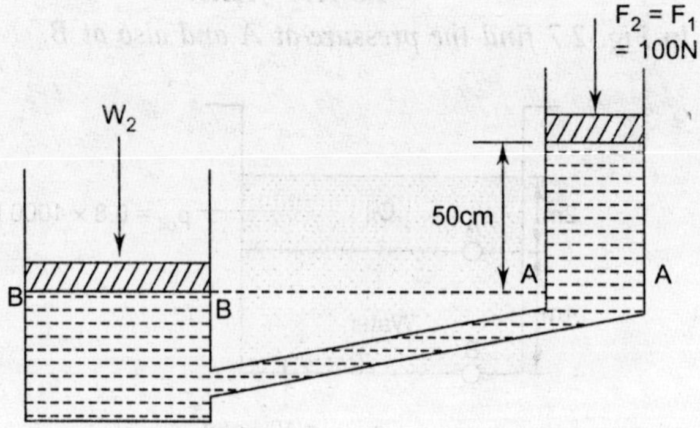

Fig. 2.9

$$W_2 = p_{AA} \times A$$
$$= 323215 \times \frac{\pi}{4}(0.08)^2 \text{ N.}$$
$$= 1624.6 \text{ N} = \text{force on large piston.} \quad \textbf{Ans.}$$

Example 2.4. *The pressure at a point in liquid is 4 N/cm². Find the corresponding height when the liquid is (i) water (ii) oil of sp. gravity 0.8.*

Solution. For water

$$z = \frac{p}{\rho g} = \frac{4 \times 10^4}{1000 \times 9.81} m$$
$$= 4.08 \text{ m} \quad \textbf{Ans.}$$

For oil

$$z = \frac{p}{\rho_{oil}\, g} = \frac{4 \times 10^4}{0.8 \times 1000 \times 9.81} = 5.1 \text{ m} \quad \textbf{Ans.}$$

Example 2.5. *An oil of specific gravity 0.8 is contained in a vessel. At a point the height of oil is 50m. Determine the corresponding water head.*

Solution

$$p = 0.8 \times 1000 \times 9.81 \times 50 \text{ N/m}^2$$

$$z_{water} = \frac{p}{\rho_{water} \times g} = \frac{0.8 \times 1000 \times 9.81 \times 50}{1000 \times 9.81} \text{ m}$$
$$= 40 \text{ m} \quad \textbf{Ans.}$$

Chapter **3**

Measurement of Pressure

3.1 INTRODUCTION

In the previous chapter we have calculated pressure by mathematical formula. Here we will measure the pressure by means of manometers and mechanical gauges.

3.2 MANOMETERS

Manometer is a device which measures pressure by balancing a column of liquid against the pressure to be measured. It can be used for measuring gauge, absolute, atmospheric and differential pressures.

3.2.1 Types of Manometers

There are mainly two types of manometers (*a*) Simple manometers (*b*) Differential manometers.

3.2.1.1 Simple Manometers

Following are in the category of simple manometers. (*i*) Piezometer, (*ii*) U-tube manometers and (*iii*) single column manometers.

(*i*) **Piezometer :** A piezometer is a vertical transparent glass tube. The lower end of it is connected to the point in the fluid container at which pressure is to be measured. The upper end is open to atmosphere (Fig. 3.1).

A and B are two gauge points.

Fluid pressure at A is P_A given by

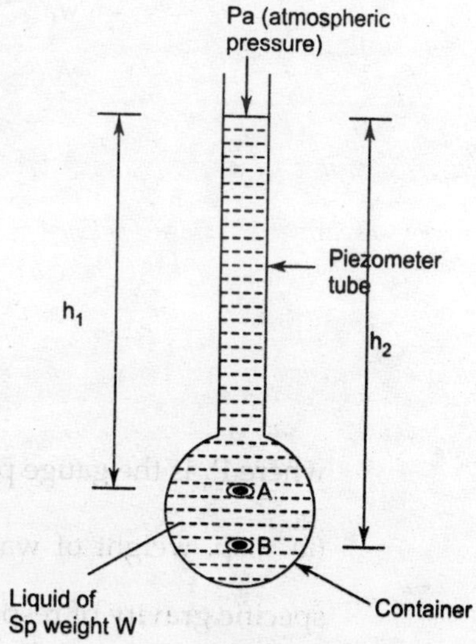

Fig. 3.1

$$P_A = h_1 w + P_a \, N/m^2 \, [h_1 \text{ in meter}, w \Rightarrow N/m^3]$$
$$[P_a = \text{atmospheric pressure}]$$

Fluid pressure at B is P_B given by

$$P_B = wh_2 + P_a, \, N/m^2$$

If P_A and P_B are to be expressed in gauge pressures, then atmospheric pressure line will be the datum *i.e.*, $P_a = 0$

So $$P_A = wh_1 \text{ and } P_B = wh_2$$

(*ii*) **U-tube manometers :** Fig. 3.2 (*a*) shows a *u*-tube double column manometer shown in an arrangement where we have to measure pressure greater than atmospheric pressure.

The pressure at A is $P_A = P_x + w_1 h_1$

Pressure at B $P_B = P_a + w_2 h_2$

$$= 0 + w_2 h_2$$
$$P_A = P_B$$
$$P_x + w_1 h_1 = w_2 h_2$$

or $$P_x = (w_2 h_2 - w_1 h_1)$$

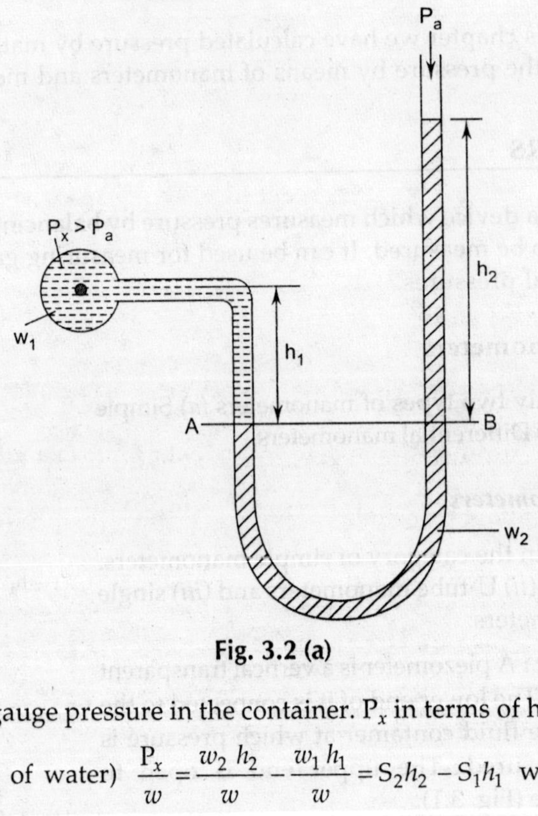

Fig. 3.2 (a)

where P_x is the gauge pressure in the container. P_x in terms of head of water column.

(w = sp. weight of water) $\dfrac{P_x}{w} = \dfrac{w_2 \, h_2}{w} - \dfrac{w_1 \, h_1}{w} = S_2 h_2 - S_1 h_1$ where S_1 & S_2 are the specific gravity of respective liquids.

Fig. 3.2 (*b*) shows a *u*-tube double column manometer shown in an arrangement where we have to measure pressure less than atmospheric pressure.

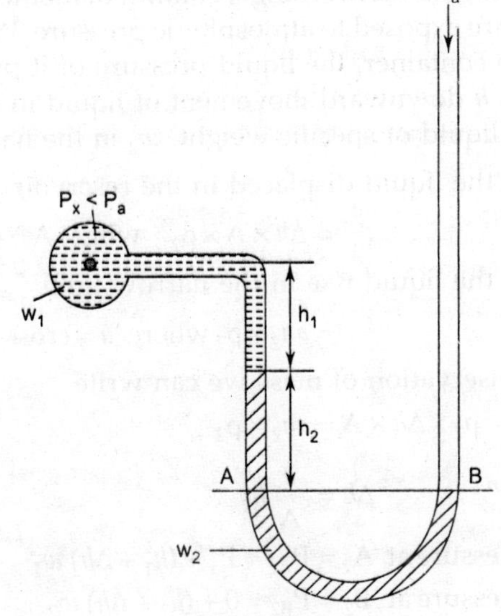

Fig. 3.2 (b)

$$P_A = w_1h_1 + w_2h_2 + P_x$$
$$P_B = P_a = 0$$
$$P_A = P_B$$

or $$w_1h_1 + w_2h_2 + P_x = 0$$

or $$P_x = -(w_1h_1 + w_2h_2) = \text{Gauge pressure in the container.}$$

In terms of head of water column

$$\frac{P_x}{w} = -\left(\frac{w_1}{w}h_1 + \frac{w_2}{w}h_2\right) = -(S_1h_1 + S_2h_2)$$

Disadvantage of double column: In double column we have to measure h_1 and h_2 and that is likely to increase the errors in the measurement. (This disadvantage is overcome by using a single column manometer).

(*iii*) **Single column manometer:** One of the limbs in double column manometer is converted into a reservoir having large cross sectional area (about 100 times) with respect to the other limb.

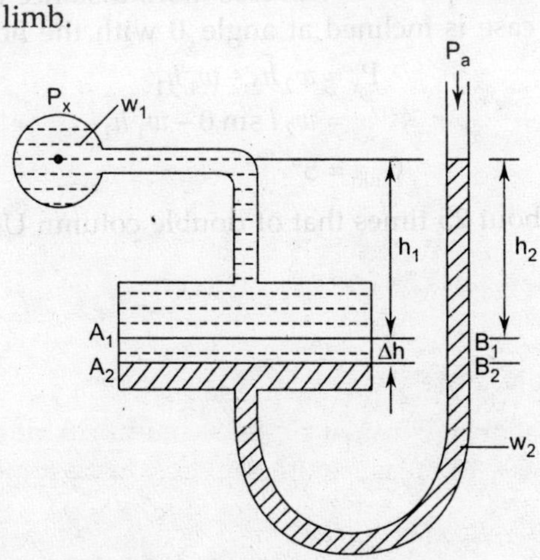

Fig. 3.3 (a)

Fig. 3.3 (*a*) shows a vertical single column manometer. Initially, the both limbs of the manometer are exposed to atmospheric pressure. When the left limb with reservoir is connected to container, the liquid pressure of it pushes down the level A_1 B_1 to A_2 B_2 causing Δh downward movement of liquid in the reservoir. Let h_2 is the rise of manometric liquid of specific weight, w_2 in the narrow limb.

The mass of the liquid displaced in the reservoir

$$= \Delta h \times A \times \rho_2 \quad \text{where A = cross-sectional area of the reservoir.}$$

The mass of the liquid rise in the narrow limb

$$= a h_2 \times \rho_2 \text{ where } a = \text{cross-sectional area of the narrow limb.}$$

From the conservation of mass we can write

$$\rho_2 \times \Delta h \times A = a h_2 \times \rho_2$$

or

$$\Delta h = \frac{a}{A} h_2$$

The gauge pressure at $A_2 = P_{A2} = P_x + (h_1 + \Delta h)\, w_1$

The gauge pressure at $B_2 = P_{B2} = 0 + (h_2 + \Delta h)\, w_2$

$$P_{A2} = P_{B2}$$

or

$$P_x + (h_1 + \Delta h)\, w_1 = (h_2 + \Delta h)\, w_2$$

or

$$P_x = (w_2 h_2 - w_1 h_1) + (w_2 - w_1)\, \Delta h$$

$$= (w_2 h_2 - w_1 h_1) + (w_2 - w_1) \frac{a}{A} h_2$$

$$= w_2 h_2 \left(1 + \frac{a}{A}\right) - w_1 \underline{\left(h_1 + h_2 \frac{a}{A}\right)}$$

Neglecting $\frac{a}{A}$, $\quad P_x = w_2 h_2 - w_1 h_1.$

If the underline term is neglected then

$\boxed{P_x = w_2 h_2,}$ so we have to measure only h_2.

Fig. 3.3 (*b*) shows an inclined manometer. This manometer is more sensitive because the manometric liquid will traverse more distance than that of vertical. The narrow limb in this case is inclined at angle θ with the horizontal.

$$P_x = w_2 h_2 - w_1 h_1$$

$$= w_2 l \sin\theta - w_1 h_1$$

$$\theta_{min} = 5°$$

Sensitivity about 25 times that of double column U-tube.

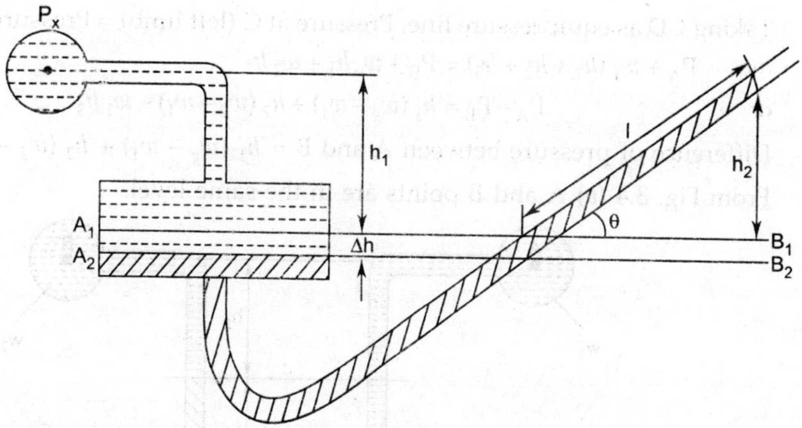

Fig. 3.3 (b)

3.2.1.2 *Differential Manometers*

Differential manometers are used to measure the pressure difference between two points in a pipe or in two different containers. There are three types of differential manometers: (*i*) U-tube differential manometers (*ii*) Inverted *u*-tube differential manometers (*iii*) Micro-manometers or *u*-tube double reservoir manometers.

(*i*) **U-tube differential manometers:** It consists of a *u*-tube, containing a heavy liquid. The two ends are connected to two different points whose pressure difference is to be measured.

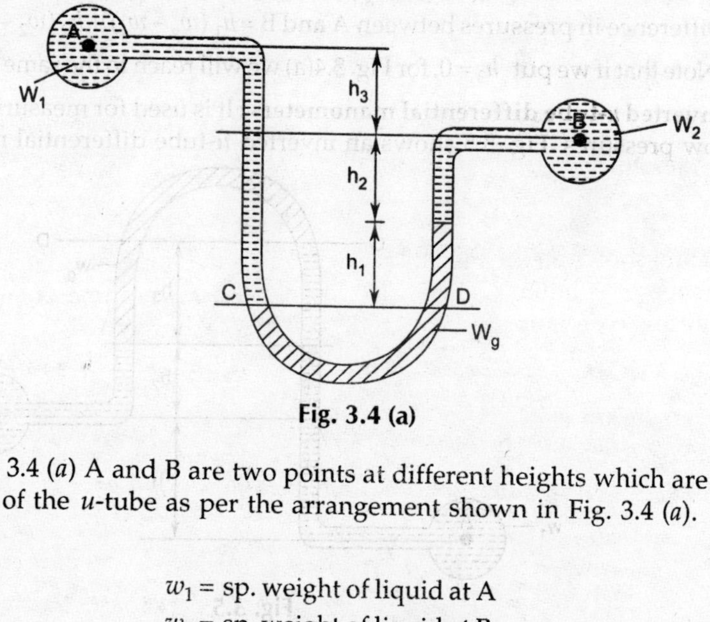

Fig. 3.4 (a)

From Fig. 3.4 (*a*) A and B are two points at different heights which are connected by two ends of the *u*-tube as per the arrangement shown in Fig. 3.4 (*a*).

Let

w_1 = sp. weight of liquid at A

w_2 = sp. weight of liquid at B

w_g = sp. weight of liquid at manometric liquid say mercury

P_A = Pressure at A & P_B = Pressure at B.

Taking CD as equipressure line, Pressure at C (left limb) = Pressure at D (right limb).

$$P_A + w_1 (h_1 + h_2 + h_3) = P_B + w_g h_1 + w_2 h_2$$

or
$$P_A - P_B = h_1 (w_g - w_1) + h_2 (w_2 - w_1) - w_1 h_3$$

Difference of pressure between A and B = $h_1 (w_g - w_1) + h_2 (w_2 - w_1) - h_3 w_1$.

From Fig. 3.4 (*b*) A and B points are in the same level.

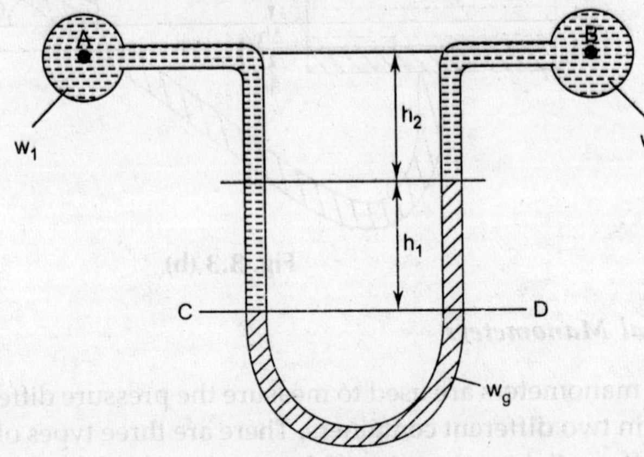

Fig. 3.4 (b)

Pressure at C (left limb) = Pressure at D (right limb)

$$P_A + (h_1 + h_2)w_1 = w_g h_1 + w_2 h_2 + P_B$$

or
$$P_A - P_B = w_g h_1 + w_2 h_2 - w_1 h_1 - w_1 h_2$$

Difference in pressures between A and B = $h_1 (w_g - w_1) + h_2 (w_2 - w_1)$

[Note that if we put $h_3 = 0$, for Fig. 3.4(a) we will reach to the same equation for Fig. 3.4(*b*)]

(*ii*) **Inverted *u*-tube differential manometers :** It is used for measuring difference between low pressures. Fig. 3.5 shows an inverted *u*-tube differential manometer.

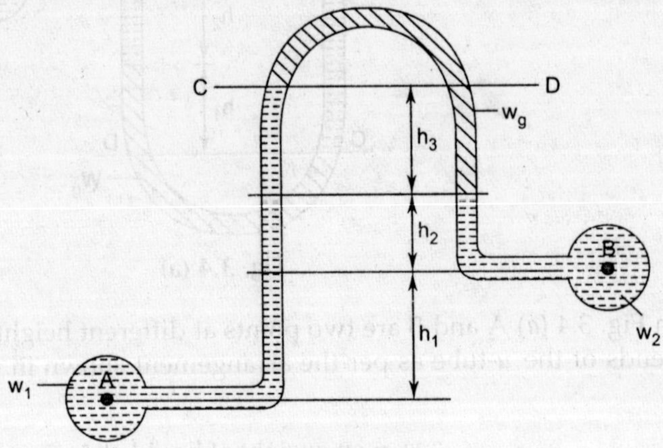

Fig. 3.5

Pressure at C (left limb) = Pressure at D (right limb)

$$P_A - w_1 (h_1 + h_2 + h_3) = P_B - w_2 h_2 - w_g h_3$$

or
$$P_A - P_B = w_1 (h_1 + h_2 + h_3) - w_2 h_2 - w_g h_3$$

Difference in pressures between A & B. $= w_1 h_1 + h_2 (w_1 - w_2) + h_3 (w_1 - w_g)$

(*iii*) **Micromanometers:** In this manometer there are two reservoirs and uses two manometric liquids of different specific gravity. Fig. 3.6 shows a micromanometer.

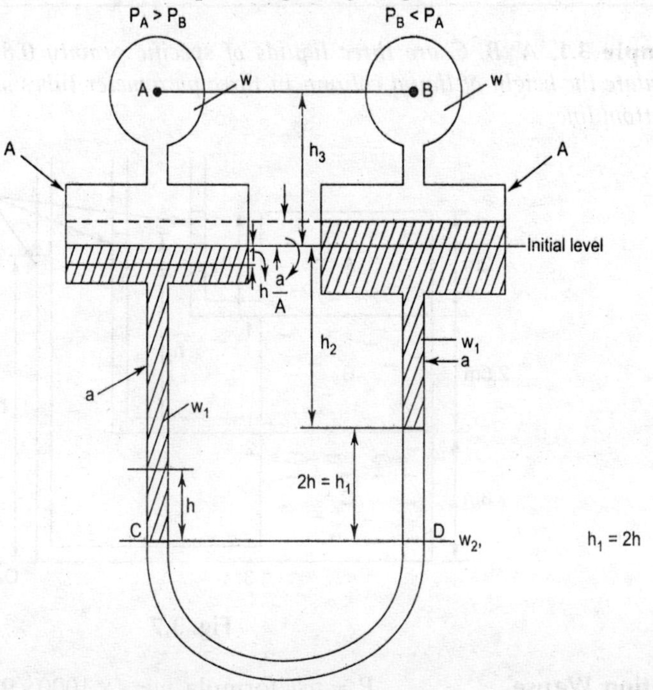

Fig. 3.6

h = fall in liquid height of sp.wt w_1 in the left limb.

The fall in reservoir level in the left limb $= h \dfrac{a}{A}$

= rise in the reservoir height in right limb.

Pressure at C = Pressure at D

$$P_A + w_1 \left(h_1 + h_2 - h\frac{a}{A} \right) + w \left(h_3 + h\frac{a}{A} \right)$$

$$= P_B + w_1 \left(h_2 + h\frac{a}{A} \right) + w_2 h_1 + w \left(h_3 - h\frac{a}{A} \right)$$

If $\qquad \dfrac{a}{A} = 0$ (neglected), then

$$P_A + w_1 (h_1 + h_2) + wh_3 = P_B + w_1 h_2 + w_2 h_1 + wh_3$$

or $\qquad P_A - P_B = w_1 h_2 + w_2 h_1 + wh_3 - w_1 h_1 - w_1 h_2 - wh_3$

$$= w_2 h_1 - w_1 h_1$$

$$= h_1 (w_2 - w_1) \qquad \text{[as } h_1 = 2h\text{]}$$

$$P_A - P_B = 2h (w_2 - w_1)$$

Difference in pressure between A & B = $2\,h\,(w_2 - w_1)$.

SOLVED EXAMPLES

Example 3.1. *A, B, C are three liquids of specific gravity 0.8, 0.85 and 0.95, respectively. Calculate the height of liquid column in three piezometer tubes shown in Fig. 3.7. Take datum as bottom line.*

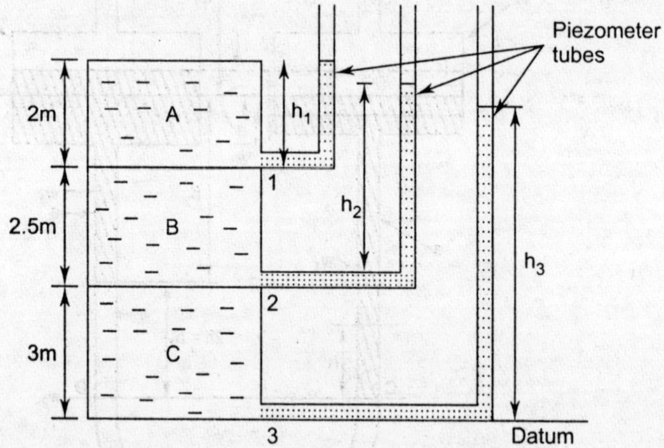

Fig. 3.7

Solution. We use $P = wh$, formula, $w = s \times 1000 \times 9.81\ \text{N}/m^3$

s = *sp.* gravity

(*i*) Pressure at 3, $P_3 = (0.95 \times 1000 \times 9.81 \times 3) + (0.85 \times 1000 \times 9.81 \times 2.5) + (0.8 \times 1000 \times 9.81 \times 2)$

or $P_3 = 64501\ \text{N}/m^2$, thus pressure head in terms of liquid C

$$= \frac{64501}{0.95 \times 1000 \times 9.81}\ m = h_3$$

or $h_3 = 6.92$ m.

(*ii*) Pressure at 2, $P_2 = (0.85 \times 1000 \times 9.81 \times 2.5) + (0.8 \times 1000 \times 9.81 \times 2)$

or $P_2 = 36542\ \text{N}/m^2$

or $$h_2 = \frac{P_2}{w_B} = \frac{36542}{0.85 \times 1000 \times 9.81} = 4.4\ m\text{(in terms of B)}$$

(*iii*) Pressure at 1, $= P_1 = 0.8 \times 1000 \times 9.81 \times 2$

$= 15696\ \text{N/m}^2$

Pressure head in terms of liquid A

$$= \frac{15696}{0.8 \times 1000 \times 9.81} = 2\ m = h_1.$$

Example 3.2. *Fig. 3.8 shows a u-tube manometer. Determine the pressure of the fluid in pipe.*

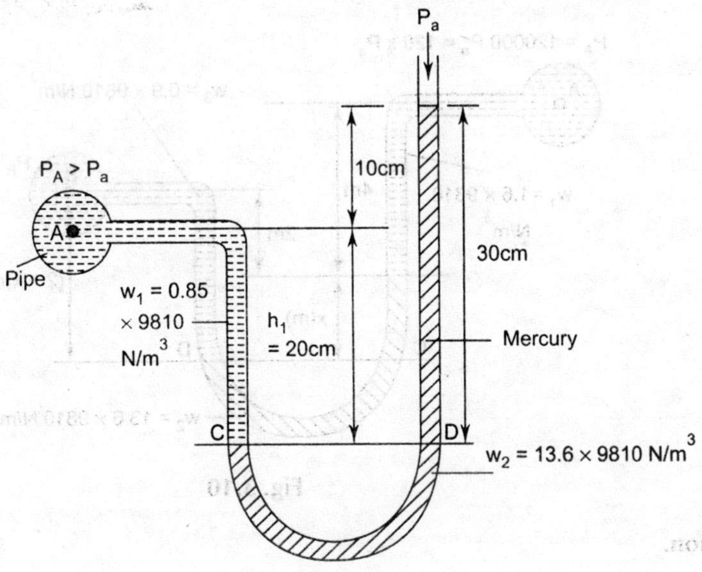

Fig. 3.8

Solution.

$$\text{Pressure at C} = \text{Pressure at D}$$

or $$P_A + 0.2 \times 0.85 \times 9810 = 0.3 \times 13.6 \times 9810$$

or $$P_A = 38357.1\ N/m^2. \quad \textbf{Ans.}$$

Example 3.3. *Fig. 3.9 shows a u-tube manometer in balanced position. Find out the vacuum pressure at A.*

Solution.

$$\text{Pressure at C} = \text{Pressure at D}$$

or $$P_A + 0.2 \times 0.9 \times 9810 + 0.5 \times 13.6 \times 9810 = 0,$$

or $$P_A = -\ 68474\ N/m^2.$$

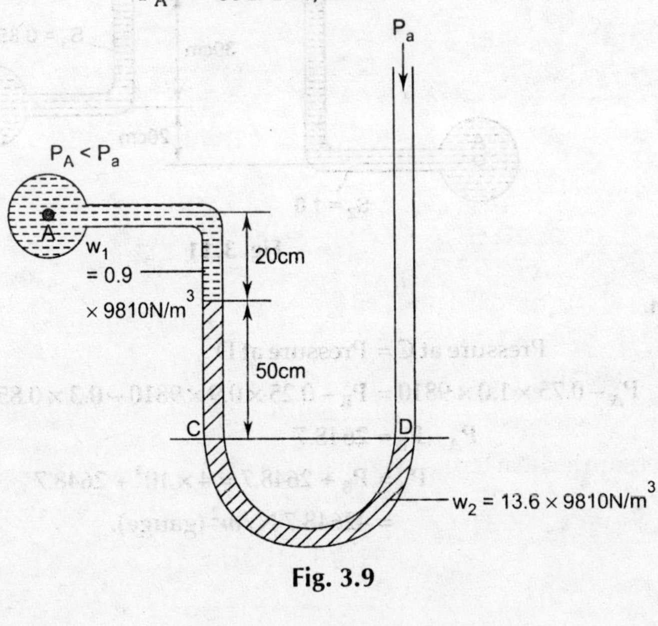

Fig. 3.9

Example 3.4. *Fig. 3.10 shows a u-tube differential manometer. Find out the value of x.*

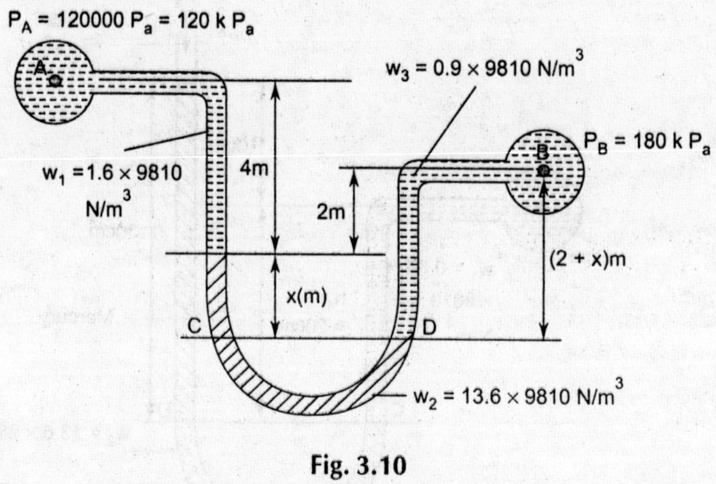

Fig. 3.10

Solution.

$$\text{Pressure at C} = \text{Pressure at D}$$

$$120 \times 10^3 + 4 \times 1.6 \times 9810 + x \times 13.6 \times 9810 = 180 \times 10^3 + (2 + x) \times 0.9 \times 9810$$

or $$x = 0.12, (m).$$

Example 3.5. *Fig. 3.11 shows an inverted u-tube differential manometer. Calculate the pressure at A if pressure at B is 4×10^4 N/m^2.*

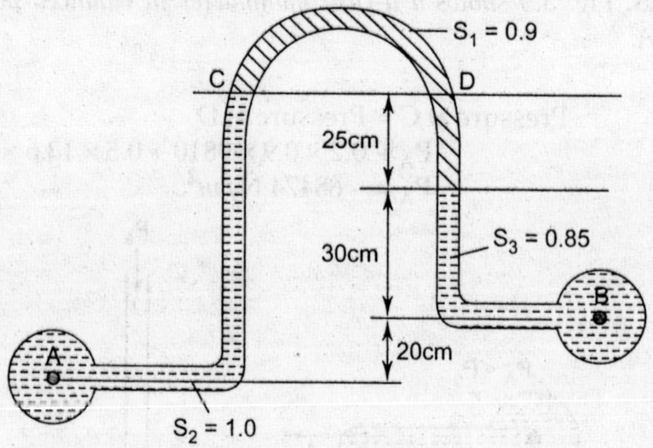

Fig. 3.11

Solution.

$$\text{Pressure at C} = \text{Pressure at D}$$

$$P_A - 0.75 \times 1.0 \times 9810 = P_B - 0.25 \times 0.9 \times 9810 - 0.3 \times 0.85 \times 9810$$

or $$P_A - P_B = 2648.7$$

or $$P_A = P_B + 2648.7 = 4 \times 10^4 + 2648.7$$

$$= 42648.7 \text{ N}/m^2 \text{ (gauge)}.$$

Example 3.6. *Fig. 3.12 shows a u-tube double reservoir manometer, find out the value of h. Given* $P_C - P_B = 1$ *cm of water.*

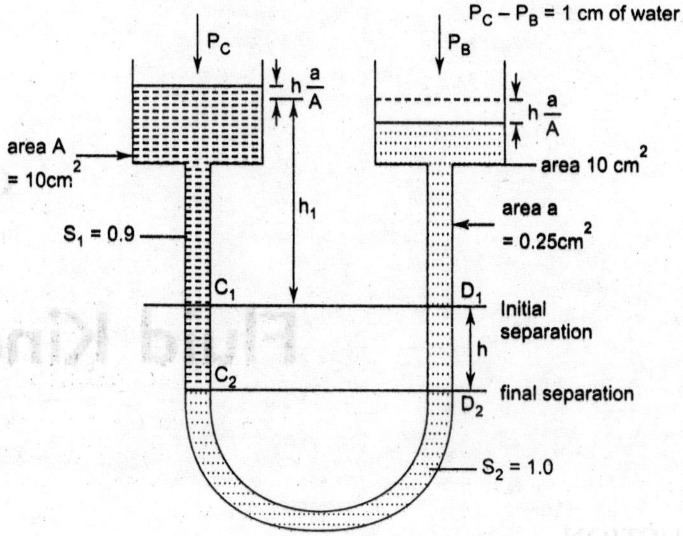

Fig. 3.12

Solution. For initial separation

$$P_B = P_C$$

$$w_1\left(h_1 + h\frac{a}{A}\right) = w_2\left(h_1 - h\frac{a}{A}\right) \quad \text{...(1)}$$

For final separation

$$\text{Pressure at } C_2 = \text{Pressure at } D_2$$

$$P_C + w_1(h_1 + h) = P_B + w_2(h_1 + h)$$

or $$P_C - P_B = w_2(h + h_1) - w_1(h + h_1)$$

or $$\frac{P_C - P_B}{w} = \frac{w_2}{w}(h + h_1) - \frac{w_1}{w}(h + h_1)$$

or $$0.01 = 1(h + h_1) - 0.9(h + h_1)$$

$$= 0.1(h + h_1)$$

or $$h + h_1 = \frac{0.01}{0.1} = \frac{1}{10} \quad \text{...(2)}$$

Again for equation (1)

$$0.9h_1 + 0.9h\frac{a}{A} = h_1 - h\frac{a}{A} \qquad \left[\text{From (1) } 0.9\left(h_1 + h\frac{a}{A}\right) = 1\left(h_1 - \frac{ha}{A}\right)\right]$$

or $$0.1h_1 = 1.9h\frac{a}{A} = 1.9\,h \times \frac{0.25}{10} = 0.0475\,h$$

or $$h_1 = 0.475\,h$$

From equation (2) we have

$$h + 0.475\,h = 0.1$$

or $$h = \frac{0.1}{1.475}\,m = 0.068\,m \quad \textbf{Ans.}$$

Chapter 4

Fluid Kinematics

4.1 INTRODUCTION

The branch of engineering science which deals with the motion of a fluid particle in terms of velocity, acceleration, time and space without considering the causes of motion (*i.e.*, force), is called fluid kinematics. In this chapter, we will find out velocity and acceleration of fluid motion.

4.2 LAGRANGIAN METHOD FOR DESCRIBING FLUID MOTION

In Lagrangian method the behaviour of each and every individual particle of fluid flow is described. The observer follows the particle and travels with the particle under study. Indentities of the particle are made by specifying their initial spatial coordinates at a given time. The position of this particle at any other instant of time is a function of its identity and time. Mathematically,

$$\vec{S} = S\left(\vec{S_o}, t\right) \quad \text{...(4.1)}$$

where $\vec{S}$ is the position vector of a particle with reference to a fixed point of reference at time t, and $\vec{S_o}$ is its initial position at a given time $t = t_o$. Thus $(\vec{S_o}, t)$ identifies the identity of the particle. Equation (4.1) can be written into scalar components of x, y, z rectangular cartesian coordinates system.

$$x = x\,(x_0, y_0, z_0, t) \quad \text{...(4.2(a))}$$

$$y = y\,(x_0, y_0, z_0, t) \quad \text{...(4.2(b))}$$

$$z = z\,(x_0, y_0, z_0, t) \quad \text{...(4.2(c))}$$

where x_0, y_0, z_0 are the initial coordinates and x, y, z are the coordinates at time t of the particle.

Now $\vec{S} = \vec{i}\,x + \vec{j}\,y + \vec{k}\,z$ where $\vec{i}$, $\vec{j}$ and $\vec{k}$ are the unit vectors along x, y and z axes respectively.

The velocity $\vec{V}$ is given by

$$\vec{V} = \left[\frac{d\vec{S}}{dt}\right]_{S_o} \quad ...(4.3)$$

If u, v and w are the scalar components of velocity in x, y and z directions respectively, we can write

$$u = \left[\frac{dx}{dt}\right]_{x_0, y_0, z_0} \quad ...(4.3(a))$$

$$v = \left[\frac{dy}{dt}\right]_{x_0, y_0, z_0} \quad ...(4.3(b))$$

$$w = \left[\frac{dz}{dt}\right]_{x_0, y_0, z_0} \quad ...(4.3(c))$$

The acceleration $\vec{a}$ is given by

$$\vec{a} = \left[\frac{d^2\vec{S}}{dt^2}\right]_{S_o} \quad ...(4.4)$$

If a_x, a_y, and a_z are the scalar components of acceleration in x, y and z directions respectively, we can write

$$a_x = \left[\frac{d^2x}{dt^2}\right]_{x_0, y_0, z_0} \quad ...(4.4(a))$$

$$a_y = \left[\frac{d^2y}{dt^2}\right]_{x_0, y_0, z_0} \quad ...(4.4(b))$$

$$a_z = \left[\frac{d^2z}{dt^2}\right]_{x_0, y_0, z_0} \quad ...(4.4(c))$$

Here (x_0, y_0, z_0) are fixed point in fluid field at $t = t_o$.

This method is rarely suitable for practical applications because the mathematical difficulties of solving equations (4.3) and (4.4).

4.3 EULERIAN METHOD FOR DESCRIBING FLUID MOTION

In Eulerian method, the velocity, acceleration, pressure, density etc., are described at a point in fluid flow. So it seeks the velocity $\vec{V}$ and its variation with time t at each and every location $(\vec{S})$ in the flow field. Mathematically,

$$\vec{V} = V(\vec{S}, t)$$

where $\vec{V} = \vec{i}\, u + \vec{j}\, v + \vec{k}\, w$ ...(4.5)

and $\vec{S} = \vec{i}\, x + \vec{j}\, y + \vec{k}\, z$ ...(4.6)

therefore $u = u(x, y, z, t)$ ...(4.6(a))

$v = v(x, y, z, t)$ ...(4.6(b))

$w = w(x, y, z, t)$ ...(4.6(c))

$$\frac{d\vec{S}}{dt} = V(\vec{S}, t) \quad \text{...(4.7)}$$

or

$$\frac{dx}{dt} = u(x, y, z, t) \quad \text{...(4.7(a))}$$

$$\frac{dy}{dt} = v(x, y, z, t) \quad \text{...(4.7(b))}$$

$$\frac{dz}{dt} = w(x, y, z, t) \quad \text{...(4.7(c))}$$

4.4 LAGRANGIAN RELATIONSHIPS FROM EULERIAN EQUATIONS

If we integrate equation (4.7), we will get the constants of integration which are found from initial condition of the fluid particle. Hence the solution of (4.7) gives the relations derived for Lagrange.

$$\vec{S} = S(\vec{S}_o, t)$$

and

$$x = x(x_0, y_0, z_0, t)$$
$$y = y(x_0, y_0, z_0, t)$$
$$z = z(x_0, y_0, z_0, t).$$

But the solution of the 3 sets of simultaneous differential equation is very difficult to reach to Lagrangian equations from Euler's equation.

Thus the Eulerian method is commonly used in fluid mechanics. Hence in this book if anything is not mentioned, it is the Eulerian method which is in use.

4.5 STEADY AND UNSTEADY FLOWS

The fluid flow will be termed as *steady* if the fluid properties and various other hydrodynamic parameters at any point do not change with time.

Thus

$$\vec{V} = V(\vec{S})$$
$$\vec{a} = a(\vec{S}).$$

The preceding equation states that in steady flow the velocity and acceleration are functions of space coordinates only. Fluid characteristics like velocity (V), pressure, (p) density (ρ) etc., at a point do not change with time. Mathematically,

$$\left[\frac{\partial V}{\partial t}\right]_{x_0, y_0, z_0} = 0, \left[\frac{\partial p}{\partial t}\right]_{x_0, y_0, z_0} = 0$$

$$\left[\frac{\partial \rho}{\partial t}\right]_{x_0, y_0, z_0} = 0$$

The fluid flow will be termed as unsteady if the fluid properties and various other hydrodynamic parameters at any point change with time.

Thus $$\left[\frac{\partial V}{\partial t}\right]_{x_0, y_0, z_0} \neq 0, \left[\frac{\partial p}{\partial t}\right]_{x_0, y_0, z_0} \neq 0, \left[\frac{\partial \rho}{\partial t}\right]_{x_0, y_0, z_0} \neq 0.$$

4.6 UNIFORM AND NON-UNIFORM FLOWS

The fluid flow will be termed as uniform if flow parameters like velocity, pressure and density etc., remain constant through out the flow field at any given time.

$$\vec{V} = V(t), p = p(t), \rho = \rho(t),$$

$$\therefore \quad \left[\frac{\partial V}{\partial s}\right]_{t=\text{const.}} = 0, \left[\frac{\partial p}{\partial s}\right]_{t=\text{const.}} = 0, \left[\frac{\partial \rho}{\partial s}\right]_{t=\text{const.}} = 0 \text{ etc.}$$

The fluid flow will be termed as non-uniform if flow parameters like velocity, pressure and density etc., change in the flow field at any given time.

$$\left[\frac{\partial V}{\partial s}\right]_{t=\text{const.}} \neq 0, \left[\frac{\partial p}{\partial s}\right]_{t=\text{const.}} \neq 0, \left[\frac{\partial \rho}{\partial s}\right]_{t=\text{const.}} \neq 0 \text{ etc.}$$

Note: Conventionally, a uniform flow is prescribed only in terms of velocity rather than in terms of all the dependent variables.

4.7 STREAM LINE

If at any fixed instant of time, the velocity vector is tangential to a imaginary space curve, in a fluid flow, the space curve so drawn is called a stream line.

$$u = \frac{dx}{dt}, v = \frac{dy}{dt}, w = \frac{dz}{dt},$$

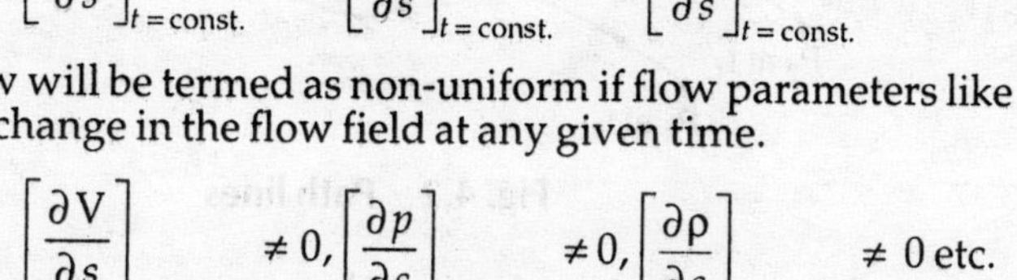

Fig. 4.1 Stream line at a steady flow

The equation of a general stream line is given by

$$\frac{dx}{u} = \frac{dy}{v} = \frac{dz}{w}, \qquad [dt \text{ is same time interval}].$$

In two rectangular coordinates in x and y, we have

$$\frac{dx}{u} = \frac{dy}{v} \text{ or } \frac{dy}{dx} = \frac{v}{u}.$$

which states that the slope of a plane stream line equals to the ratio of their respective velocity.

4.8 PATH LINES

The path lines are the lines or trajectory of different fluid particles traversed by them at different points of time. Path lines are the outcome of Lagrangian method in describing fluid flow.

Let P_1, P_2 and P_3 are the different fluid particle in a fluid flow. The paths traversed by them are shown in Fig. 4.2.

One path can intersect the other.

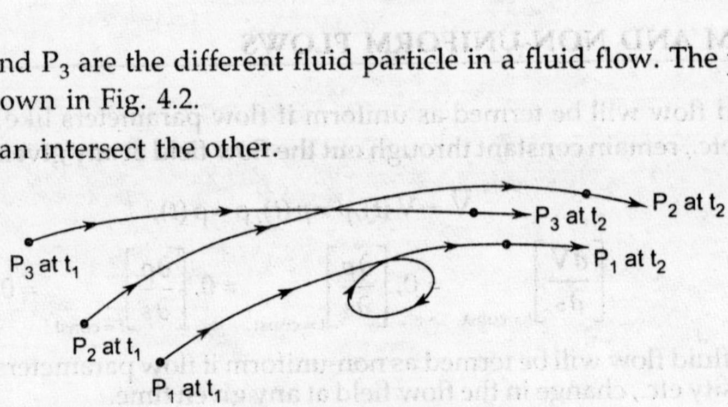

Fig. 4.2 Path lines

The path traced by a single particle of smoke coming out from a cigarette is a path line.

Note: For steady flow the path lines are identical to stream lines.

4.9 STREAK LINES

Let A be a fixed point in space, at time t_1. P_1, P_2 and P_3 (all passing through A) are the three path lines traversed by three particles in an unsteady fluid flow. Let B, C and D are the end points at time t_2 of P_1, P_2, P_3 respectively. If we join A B C D, we will get a line which is called the streak line. Thus a streak line at any instant of time is the locus of the temporary locations of all particles that have passed through a fixed point in the flow field.

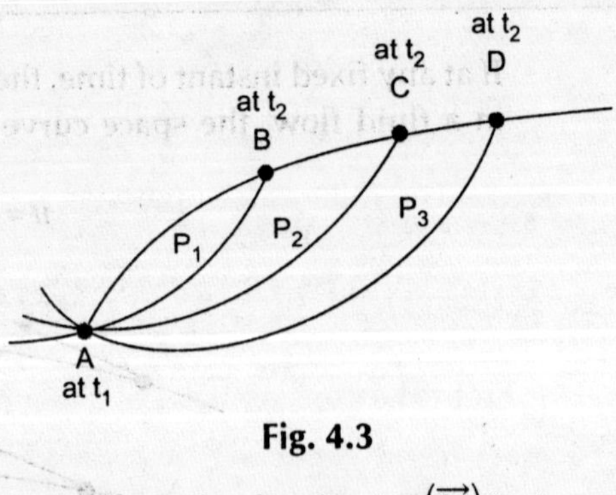

Fig. 4.3

In Lagrangian method, if a fluid particle $(\vec{S_0})$ passes through a fixed point $(\vec{S_1})$ in time t, then $$S(\vec{S_0}, t) = \vec{S_1} \quad ...(4.8)$$

The equation of streak line at a time t is given by

$$\vec{S} = f(\vec{S_0}, t) \quad ...(4.9)$$

$$= f[F(\vec{S_1}, t), t] \quad [\text{where } \vec{S_0} = F(\vec{S_1}, t) \text{ derived from (4.8)}]$$

or $$\vec{S} = f[F(\vec{S_1}, t), t] \quad ...(4.10)$$

Equation (4.10) is the equation for a streak line referrent to a fixed point $(\vec{S_1})$.

4.10 ACCELERATION OF A FLUID PARTICLE

We know
$$u = u(x, y, z, t)$$
$$v = v(x, y, z, t)$$
$$w = w(x, y, z, t)$$

or $$a_x = \frac{du}{dt} = \frac{\partial u}{\partial x}.\frac{dx}{dt} + \frac{\partial u}{\partial y}.\frac{dy}{dt} + \frac{\partial u}{\partial z}.\frac{dz}{dt} + \frac{\partial u}{\partial t} \quad \text{[By chain rule]}$$

or $$a_x = u\frac{\partial u}{\partial x} + v\frac{\partial u}{\partial y} + w\frac{\partial u}{\partial z} + \frac{\partial u}{\partial t} \quad ...(4.11)$$

where $$u = \frac{dx}{dt}, v = \frac{dy}{dt} \text{ and } w = \frac{dz}{dt}$$

Similarly

$$a_y = su\frac{\partial v}{\partial x} + v\frac{\partial v}{\partial y} + w\frac{\partial v}{\partial z} + \frac{\partial v}{\partial t} \quad ...(4.12)$$

and $$a_z = u\frac{\partial w}{\partial x} + v\frac{\partial w}{\partial y} + w\frac{\partial w}{\partial z} + \frac{\partial w}{\partial t} \quad ...(4.13)$$

Equations (4.11), (4.12), (4.13) indicate that acceleration can result from two things, one from terms $\frac{\partial u}{\partial t}, \frac{\partial v}{\partial t}$ and $\frac{\partial w}{\partial t}$ which are called *temporal* or *local* acceleration; another from terms $u\frac{\partial u}{\partial x} + v\frac{\partial u}{\partial y} + w\frac{\partial u}{\partial z}, u\frac{\partial v}{\partial x} + v\frac{\partial v}{\partial y} + w\frac{\partial v}{\partial z}$ and $u\frac{\partial w}{\partial x} + v\frac{\partial w}{\partial y} + w\frac{\partial w}{\partial z}$ in which each term $\left(u\frac{\partial u}{\partial x} \text{ etc.}\right)$ represents *convective* acceleration.

The total accelerations a_x, a_y and a_z are called material or substantial acceleration.

4.11 CONTINUITY EQUATION

From the conservation of mass in a closed system we can write down

$$\Delta m_1 = \Delta m_2$$

or $$A_1 V_1 \rho_1 \Delta t = A_2 V_2 \rho_2 \Delta t$$

or $$A_1 V_1 \rho_1 = A_2 V_2 \rho_2 \quad ...(4.14)$$

A, V, ρ represent area, velocity and density. Subscripts 1 & 2 are used to indicate the parameters at section (1 – 1) and (2 – 2).

Equation (4.14) is known as continuity equation applicable to both compressible and incompressible fluids.

For incompressible fluid

$$\rho_1 = \rho_2$$

Equation (4.14) reduces to

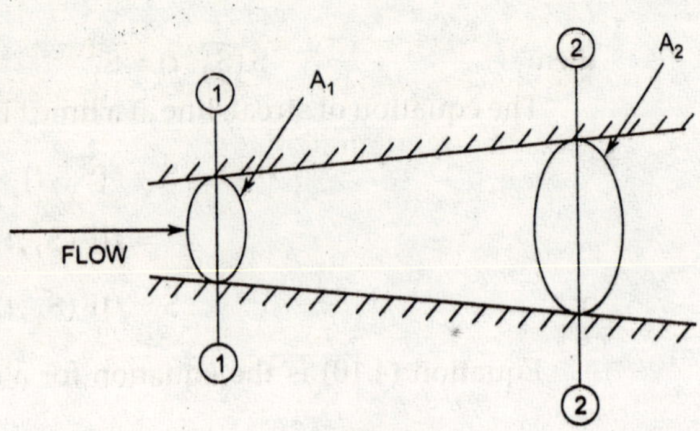

Fig. 4.4

$$\boxed{A_1V_1 = A_2V_2} \quad ...(4.15)$$

4.12 CONTINUITY EQUATION IN THREE DIMENSIONS IN A DIFFERENTIAL FORM

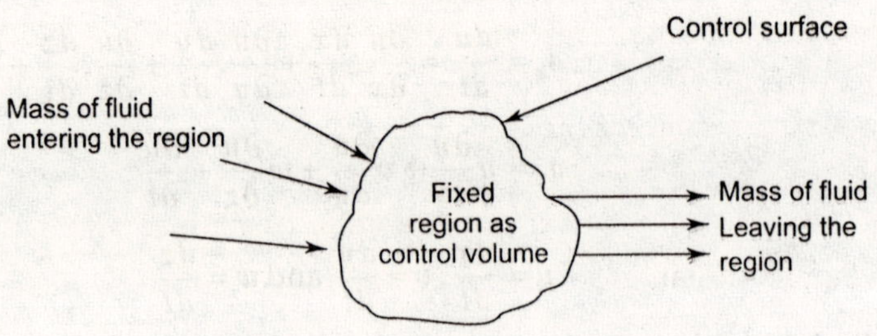

Fig. 4.5

Fig. 4.5 explains the concept of control volume and control surface. Fig. 4.6 shows an elemental rectangular parallelopiped of fluid. The volume (control) is *dx dy dz*.

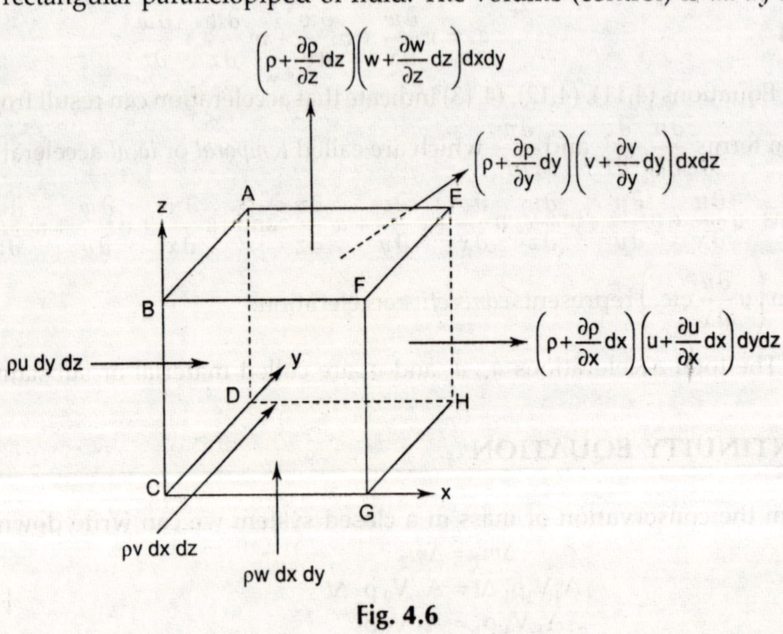

Fig. 4.6

We have rate at which mass enters the control volume = Rate at which mass leaves the control volume + rate of accumulation of mass in the control volume ...(4.16)

Thus in x direction

Rate of mass entering surface ABCD = $\rho u\, dy\, dz$

Rate of mass leaving surface

$$\text{EFGH} = \left(\rho + \frac{\partial \rho}{\partial x} dx\right)\left(u + \frac{\partial u}{\partial x} dx\right) dy\, dz$$

$$= \left[\rho u + \rho \frac{\partial u}{\partial x} dx + u \frac{\partial \rho}{\partial x} dx + \frac{\partial \rho}{\partial x} \cdot \frac{\partial u}{\partial x} (dx)^2\right] dy\, dz$$

$$= \left[\rho u + \frac{\partial(\rho u)}{\partial x} dx\right] dy\, dz, \qquad \left[\frac{\partial \rho}{\partial x} \frac{\partial u}{\partial x} (dx)^2 = 0 \text{ neglected}\right]$$

From equation (4.16) we can write down

$$\rho u\, dy\, dz = \left[\rho u + \frac{\partial(\rho u)}{\partial x} dx\right] dy\, dz + \text{Rate of accumulation of mass in control volume in } x \text{ direction.}$$

Rate of accumulation of mass in the control volume in x direction

$$= \rho u\, dy\, dz - \left[\rho u + \frac{\partial(\rho u)}{\partial x} dx\right] dy\, dz$$

$$= -\frac{\partial(\rho u)}{\partial x} dx\, dy\, dz \qquad \text{...(4.17)}$$

Similarly, Rate of accumulation of mass in the control volume in y direction.

$$= -\frac{\partial(\rho v)}{\partial y} dx\, dy\, dz \qquad \text{...(4.18)}$$

and rate of accumulation of mass in the control volume in z direction.

$$= -\frac{\partial(\rho w)}{\partial z} dx\, dy\, dz \qquad \text{...(4.19)}$$

Adding equations (4.17), (4.18) and (4.19) we get

Total rate of accumulation in the rectangular parallelopiped

$$= -\left[\frac{\partial(\rho u)}{\partial x} + \frac{\partial(\rho v)}{\partial y} + \frac{\partial(\rho w)}{\partial z}\right] dx\, dy\, dz.$$

$$\text{Thus} \quad \frac{\partial(\rho\,dx\,dy\,dz)}{\partial t} = -\left[\frac{\partial(\rho u)}{\partial x} + \frac{\partial(\rho v)}{\partial y} + \frac{\partial(\rho w)}{\partial z}\right] dx\,dy\,dz$$

[as $dx\,dy\,dz$ = control volume = constant]

or
$$\frac{\partial \rho}{\partial t} = -\left[\frac{\partial(\rho u)}{\partial x} + \frac{\partial(\rho v)}{\partial y} + \frac{\partial(\rho w)}{\partial z}\right]$$

or
$$\frac{\partial \rho}{\partial t} + \frac{\partial(\rho u)}{\partial x} + \frac{\partial(\rho v)}{\partial y} + \frac{\partial(\rho w)}{\partial z} = 0 \qquad \text{...(4.20)}$$

Equation (4.20) is the general equation of continuity. It is applicable to

(*i*) steady and unsteady flow
(*ii*) uniform and non-uniform flow, and
(*iii*) compressible and incompressible flow.

Case (*a*) Steady flow :

For steady flow $\frac{\partial \rho}{\partial t} = 0$, the continuity equation becomes

$$\frac{\partial(\rho u)}{\partial x} + \frac{\partial(\rho v)}{\partial y} + \frac{\partial(\rho w)}{\partial t} = 0 \qquad \text{...(4.21)}$$

Case (*b*) Incompressible steady flow :

For incompressible steady flow $\frac{\partial \rho}{\partial t} = 0$ and ρ = constant, the continuity equation becomes

$$\frac{\partial u}{\partial x} + \frac{\partial v}{\partial y} + \frac{\partial w}{\partial z} = 0 \qquad \text{...(4.22)}$$

Equation (4.20) can be written in a vector form

$$\frac{\partial \rho}{\partial t} + \overline{\nabla}.\left(\rho \overline{V}\right) = 0$$

where $\vec{V}$ represents the velocity vector and Laplace operator

$$\left[\vec{\nabla} = \vec{i}\frac{\partial}{\partial x} + \vec{j}\frac{\partial}{\partial y} + \vec{k}\frac{\partial}{\partial z}\right].$$

Equation (4.21) can be written in a vector form

$$\vec{\nabla}.(\rho\,\vec{V}) = 0$$

and equation (4.22) can be written in a vector form also as

$$\vec{\nabla}.(\vec{V}) = 0.$$

4.13 CONTINUITY EQUATION IN A CYLINDRICAL POLAR COORDINATE SYSTEM

Let us consider the elementary control volume with respect to (r, θ, and z) coordinates system. Fig. 4.7 is self explanatory.

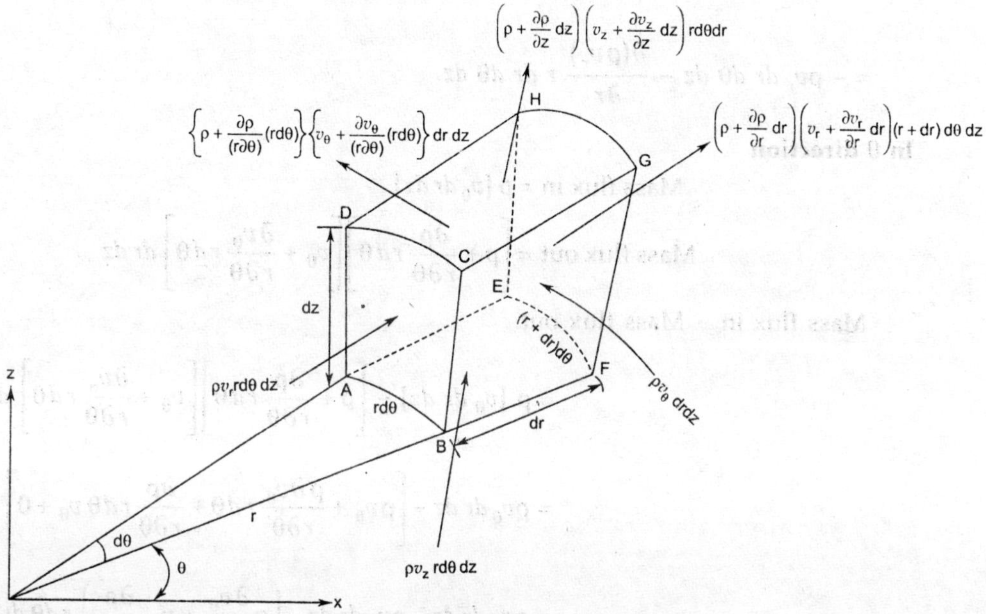

Fig. 4.7

In r direction

$$\text{Mass flux in} = \rho\, [v_r\, r\, d\theta\, dz].$$

$$\text{Mass flux out} = \left(\rho + \frac{\partial \rho}{\partial r} dr\right)\left[v_r + \frac{\partial v_r}{\partial r} dr\right](r + dr)\, d\theta\, dz$$

$\therefore$ Mass flux in – Mass flux out

$$= \rho\, [v_r . r d\theta\, dz] - \left(\rho + \frac{\partial \rho}{\partial r} dr\right)\left[v_r + \frac{\partial v_r}{\partial r} dr\right](r + dr)\, d\theta\, dz$$

$$= \rho\, [v_r\, r d\theta\, dz] - \left[\rho v_r + \frac{\rho \partial v_r}{\partial r} dr + v_r \frac{\partial \rho}{\partial r} dr + \frac{\partial \rho}{\partial r} . \frac{\partial v_r}{\partial r}(dr)^2\right][r d\theta\, dz + dr\, d\theta\, dz].$$

Neglecting term containing $(dr)^2$ *i.e.*, higher order.

$$= \rho v_r\, r d\theta\, dz - \left[\rho v_r + \frac{\rho \partial v_r}{\partial r} dr + v_r \frac{\partial \rho}{\partial r} dr + 0\right][r d\theta\, dz + dr\, d\theta\, dz]$$

$$= \rho v_r\, r d\theta\, dz - \left[\rho r v_r\, d\theta dz + \frac{\rho \partial v_r}{\partial r} r d\theta dz dr + v_r \frac{\partial \rho}{\partial r} r d\theta dr dz + \rho v_r\, dr d\theta dz + 0 + 0\right]$$

$$= \rho v_r\, r d\theta\, dz - \rho r\, v_r\, d\theta\, dz - \rho \frac{\partial v_r}{\partial r} r\, d\theta\, dz\, dr - v_r \frac{\partial \rho}{\partial r} r\, d\theta\, dz\, dr - \rho\, v_r\, dr\, d\theta\, dz$$

$$= -\rho v_r\, dr\, d\theta\, dz - \left(\rho \frac{\partial v_r}{\partial r} + v_r \frac{\partial \rho}{\partial r}\right) r d\theta\, dr\, dz$$

$$= -\rho v_r\, dr\, d\theta\, dz - \frac{\partial(\rho v_r)}{\partial r} r\, dr\, d\theta\, dz.$$

In θ direction

$$\text{Mass flux in} = \rho\, [v_\theta\, dr\, dz]$$

$$\text{Mass flux out} = \left[\rho + \frac{\partial \rho}{r\partial\theta} r\, d\theta\right]\left[v_\theta + \frac{\partial v_\theta}{r\partial\theta} r\, d\theta\right] dr\, dz$$

Mass flux in – Mass flux out

$$= \rho\, [v_\theta\, dr\, dz] - \left[\rho + \frac{\partial \rho}{r\partial\theta} r\, d\theta\right]\left[v_\theta + \frac{\partial v_\theta}{r\partial\theta} r\, d\theta\right] dr\, dz$$

$$= \rho v_\theta\, dr\, dz - \left[\rho v_\theta + \frac{\rho\, \partial v_\theta}{r\partial\theta} r\, d\theta + \frac{\partial \rho}{r\partial\theta} r\, d\theta\, v_\theta + 0\right] dr\, dz$$

$$= \rho v_\theta\, dr\, dz - \rho v_\theta\, dr\, dz - \left(\rho \frac{\partial v_\theta}{r\partial\theta} + v_\theta \frac{\partial \rho}{r\partial\theta}\right) r\, d\theta\, dr\, dz$$

$$= -\frac{\partial(\rho v_\theta)}{r\partial\theta} r\, dr\, d\theta\, dz.$$

In Z direction

$$\text{Mass flux in} = \rho\, [v_z\, r d\theta\, dr]$$

$$\text{Mass flux out} = \left[\rho + \frac{\partial \rho}{\partial z} dz\right]\left[v_z + \frac{\partial v_z}{\partial z} dz\right] r\, d\theta\, dr$$

∴ Mass flux in – Mass flux out

$$= \rho\, [v_z\, r\, d\theta\, dr] - \left[\rho + \frac{\partial \rho}{\partial z} dz\right]\left[v_z + \frac{\partial v_z}{\partial z} dz\right] r\, d\theta\, dr$$

$$= \rho\, v_z\, r\, d\theta\, dr - \left[\rho v_z + \rho \frac{\partial v_z}{\partial z} dz + v_z \frac{\partial \rho}{\partial z} dz + 0\right] r\, d\theta\, dr$$

$$= \rho\, v_z\, r\, d\theta\, dr - \rho\, v_z\, r\, d\theta\, dr - \left(\rho \frac{\partial v_z}{\partial z} + v_z \frac{\partial \rho}{\partial z}\right) r\, d\theta\, dr\, dz$$

$$= -\frac{\partial(\rho v_z)}{\partial z}\, r\, d\theta\, dr\, dz.$$

Total mass flux accumulation in r, θ and z directions

$$= -\rho v_r\, dr\, d\theta\, dz - \frac{\partial(\rho v_r)}{\partial r}\, r\, dr\, d\theta\, dz - \frac{\partial(\rho v_\theta)}{r\partial\theta}\, r\, dr\, d\theta\, dz - \frac{\partial(\rho v_z)}{\partial z}\, r\, dr\, d\theta\, dz.$$

But we have the mass flux accumulation $= \frac{\partial}{\partial t}(\rho\, r\, dr\, d\theta\, dz)$

$$= \frac{\partial\rho}{\partial t}(r\, dr\, d\theta\, dz)$$

Equating both we get,

$$-\rho v_r\, dr\, d\theta\, dz - \frac{\partial(\rho v_r)}{\partial r} r\, dr\, d\theta\, dz - \frac{\partial(\rho v_\theta)}{r\partial\theta} r\, dr\, d\theta\, dz - \frac{\partial(\rho v_z)}{\partial z} r\, dr\, d\theta\, dz = \frac{\partial\rho}{\partial t}(r\, dr\, d\theta\, dz).$$

Dividing both sides by $r\, dr\, d\theta\, dz$ we get,

$$-\frac{\rho v_r}{r} - \frac{\partial(\rho v_r)}{\partial r} - \frac{\partial(\rho v_\theta)}{r\partial\theta} - \frac{\partial(\rho v_z)}{\partial z} = \frac{\partial\rho}{\partial t}$$

or,
$$\frac{\partial(\rho v_r)}{\partial r} + \frac{1}{r}\cdot\frac{\partial(\rho v_\theta)}{\partial\theta} + \frac{\partial(\rho v_z)}{\partial z} + \frac{\rho v_r}{r} + \frac{\partial\rho}{\partial t} = 0 \qquad \text{...(4.23)}$$

Equation (4.23) is 3-D continuity equation in r, θ, z coordinates for compressible and unsteady flow.

Different forms of 2-D continuity equation under different flow condition (r, θ coordinates)

(1) 2 - D Compressible unsteady

$$\frac{\partial(\rho v_r)}{\partial r} + \frac{1}{r}\frac{\partial(\rho v_\theta)}{\partial\theta} + \rho\frac{vr}{r} + \frac{\partial\rho}{\partial t} = 0$$

(*i.e.*, there is no term involving z).

(2) 2 - D compressible steady

$$\frac{\partial(\rho v_r)}{\partial r} + \frac{1}{r}\cdot\frac{\partial(\rho v_\theta)}{\partial\theta} + \frac{\rho v_r}{r} = 0 \qquad \left[\text{as } \frac{\partial\rho}{\partial t} = 0 \text{ for steady flow}\right]$$

(3) 2 - D Incompressible unsteady

$$\frac{\partial(v_r)}{\partial r} + \frac{1}{r}\frac{\partial v_\theta}{\partial\theta} + \frac{v_r}{r} = 0 \qquad \left[\begin{array}{l}\text{for incompressible } \rho = \text{contant} \\ \therefore \frac{\partial\rho}{\partial t} = 0\end{array}\right]$$

(4) 2 - D Incompressible steady

$$\frac{\partial (v_r)}{\partial r} + \frac{1}{r} \cdot \frac{\partial v_\theta}{\partial \theta} + \frac{v_r}{r} = 0.$$

SOLVED EXAMPLES

Example 4.1. *An incompressible fluid flows through a pipe whose diameters at section 1 and 2 are 15 cm and 20 cm. Determine the discharge through the pipe if the velocity of fluid at section (1) is 10 m/s. Find also the velocity at section (2).*

Solution: $A_1V_1 = A_2V_2$

$$\frac{\pi}{4}(0.15)^2 \times 10 = \frac{\pi}{4}(0.20)^2 \times V_2$$

or $V_2 = 5.625 \text{ m/s}$

= velocity at section (2) **Ans.**

$$\text{Discharge } Q = A_1V_1 = \frac{\pi}{4}(0.15)^2 \times 10$$

$$= 0.1767 \text{ m}^3/\text{s}$$ **Ans.**

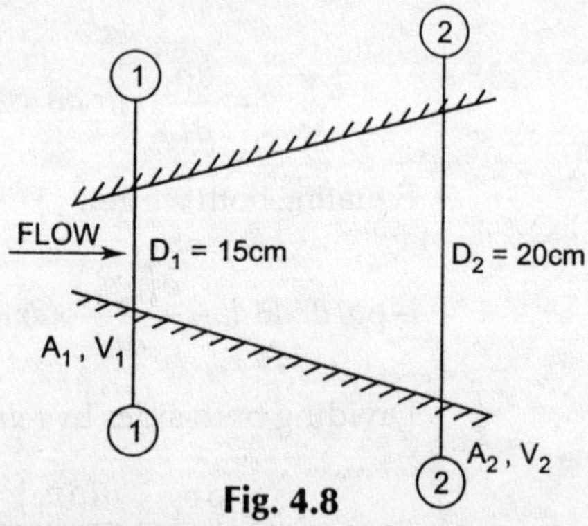

Fig. 4.8

Example 4.2. *A pipe PQ branches into two pipes QR and QS as shown in Fig. 4.9. The pipe PQ, QR and QS are having diameters of 50 cm, 40 cm and 30 cm respectively. Find out the discharge through PQ if the velocity at P is 4 m/s. Also determine the velocities at Q and S, if the velocity at R is 5 m/s.*

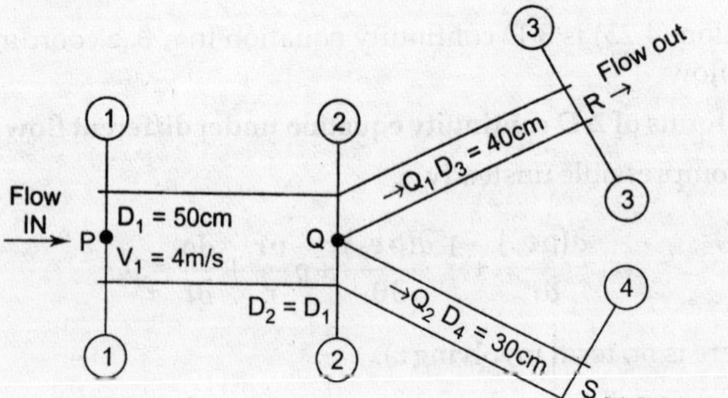

Fig. 4.9

Solution.

$$\text{Discharge through PQ} = A_1V_1 = \frac{\pi}{4}(0.5)^2 \times 4 \text{ m}^3/\text{s}$$

$$Q = 0.785 \text{ m}^3/\text{s}.$$ **Ans.**

$$Q = Q_1 + Q_2$$

$$0.785 = A_3V_3 + A_4V_4$$

$$= \frac{\pi}{4}(0.4)^2 \times 5 + \frac{\pi}{4}(0.3)^2 \times V_4$$

or $\quad V_4 = 2.216 \text{ m/s} = \text{velocity at s.}$ **Ans.**

As $\quad A_1 = A_2$, Thus $V_1 = V_2 = 4$ m/s. **Ans.**

Example 4.3. *The velocity of a fluid is given by*

$$\vec{V} = 3x^2\vec{i} - 8x^2y\vec{j} + 4t\vec{k},$$

find out the velocity and acceleration of the fluid at point (2, 1, 3) at t = 1.

Solution. $\quad u = 3x^2 = 3(2)^2 = 12$ units (x component)

$v = -8x^2y = -8(2)^2 1 = -32$ units (y component)

$w = 4t = 4 \times 1 = 4$ units (z component)

Thus $\quad \vec{V} = 12\vec{i} - 32\vec{j} + 4\vec{k},$

$$|V| = \sqrt{12^2 + (-32)^2 + 4^2} = 34.41 \text{ units.} \quad \textbf{Ans.}$$

Acceleration

$$a_x = u\frac{\partial u}{\partial x} + v\frac{\partial u}{\partial y} + w\frac{\partial u}{\partial z} + \frac{\partial u}{\partial t}$$

Now $\quad \frac{\partial u}{\partial x} = 6x = 6 \times 2 = 12$

$$\frac{\partial u}{\partial y} = 0$$

$$\frac{\partial u}{\partial z} = 0$$

$$\frac{\partial u}{\partial t} = 0$$

$\therefore \quad a_x = 12 \times 12 = 144$ units. **Ans.**

$$a_y = u\frac{\partial v}{\partial x} + v\frac{\partial v}{\partial y} + w\frac{\partial v}{\partial z} + \frac{\partial v}{\partial t}$$

$$= 12 \times (-16xy) + (-32)(-8x^2) + 4(0) + 0$$

$$= 12 \times (-16 \times 2 \times 1) + (-32)(-8 \times 2^2)$$

$$= -384 + 1024 = 640 \text{ units.} \quad \textbf{Ans.}$$

$$a_z = u\frac{\partial w}{\partial x} + v\frac{\partial w}{\partial y} + w\frac{\partial w}{\partial z} + \frac{\partial w}{\partial t}$$

$$= 12(0) + v(0) + w(0) + 4$$

$$= 4 \text{ units.} \quad \textbf{Ans.}$$

Example 4.4. *A fluid flow is defined as*

(i) $u = -4\frac{x}{y}, v = 4\ln(xy)$ *(ii)* $u = 2x^2 - xy, v = x^2 - 4xy + y^2$; *which of (i) and (ii) satisfies the steady, incompressible and 2 - D flow.*

Solution. (*i*)

$$u = -4\frac{x}{y}, v = 4\ln(xy)$$

or

$$\frac{\partial u}{\partial x} = -\frac{4}{y}$$

$$\frac{\partial v}{\partial y} = \frac{4}{xy} \times x = \frac{4}{y}$$

or

$$\frac{\partial u}{\partial x} + \frac{\partial v}{\partial y} = -\frac{4}{y} + \frac{4}{y} = 0$$

The continuity equation is satisfied and hence (*i*) satisfies the condition given (flow is possible).

(*ii*)

$$u = 2x^2 - xy, v = x^2 - 4xy + y^2$$

$\therefore$

$$\frac{\partial u}{\partial x} = 4x - y, \frac{\partial v}{\partial y} = -4x + 2y$$

$\therefore$

$$\frac{\partial u}{\partial x} + \frac{\partial v}{\partial y} = 4x - y - 4x + 2y = y \neq 0$$

So (*ii*) does not satisfy (flow is not possible).

Chapter **5**

Dynamics of Ideal Fluids

5.1 INTRODUCTION

The dynamic behaviours of fluid are studied considering the cause of the motion *i.e.*, force. We will use Newton's 2nd law of motion to derive differential equation for dynamic behaviour of ideal fluid (ρ = constant & $\mu = 0$).

5.2 BERNOULLI'S EQUATION

There are many forces which act on a fluid element. They are

F_g = gravity force,

F_p = pressure force,

F_v = viscous force,

F_t = turbulence force,

and F_c = compressibility force.

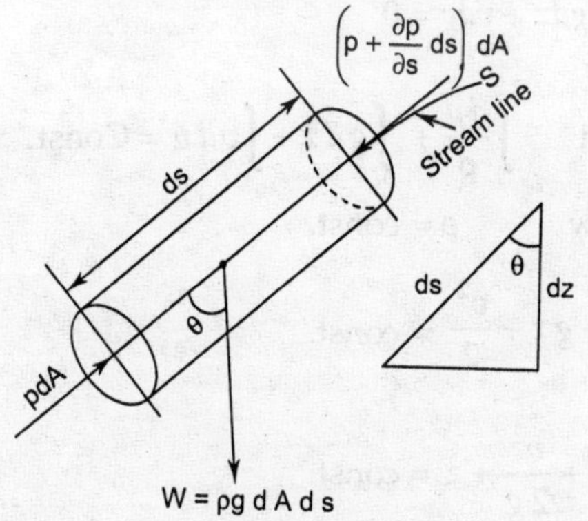

Fig. 5.1

Here we will consider gravity and pressure force to derive Bernoulli's equation.

Fig. 5.1 shows dynamic equilibrium of fluid element under the action of forces as shown in the figure. Here p denotes for intensity of pressure. By using Newton's 2nd law we get

$$\Sigma \text{ forces in s direction} = \text{mass} \times \text{acceleration in s direction}$$

$$pdA - \left(p + \frac{\partial p}{\partial s} ds\right) dA - \rho g dA\, ds \cos\theta$$

$$= \rho dA\, ds \times a_s \qquad ...(5.1)$$

where a_s = acceleration in s direction

$$v = v(s, t)$$

$$a_s = \frac{dv}{dt} = \frac{\partial v}{\partial s}\frac{ds}{dt} + \frac{\partial v}{\partial t}$$

$$= v.\frac{\partial v}{\partial s} + \frac{\partial v}{\partial t}$$

$$= v\frac{\partial v}{\partial s}, \text{ for steady flow } \frac{\partial v}{\partial t} = 0$$

From (5.1)

$$-\frac{\partial p}{\partial s} ds dA - \rho g dA\, dz = \rho dA ds \times v\frac{\partial v}{\partial s}$$

or
$$-\frac{\partial p}{\partial s} - \rho g\frac{dz}{ds} = \rho v\frac{\partial v}{\partial s} \qquad [ds\cos\theta = dz]$$

or
$$\frac{\partial p}{\rho} + gdz + vdv = 0 \qquad ...(5.2)$$

Equation (5.2) is the **Euler's equation of motion.** For steady, incompressible flow we can write down

$$\frac{\partial p}{\rho} + gdz + vdv = 0$$

Integrating we get $\int \frac{dp}{\rho} + \int g\,dz + \int v\,dv = \text{Const.}$

or incompressible flow ρ = const.

or
$$\frac{p}{\rho} + gz + \frac{v^2}{2} = \text{const.}$$

or
$$\frac{p}{\rho g} + \frac{v^2}{2g} + z = \text{const.} \qquad ...(5.3)$$

Equation (5.3) is known as **Bernoulli's equation.**

5.3 TOTAL HEAD

Each term of the Bernoulli's equation represent the dimension of length *i.e.*, in S.I. it is meter. Left hand side *i.e.*, $\left(\frac{p}{\rho g}+\frac{v^2}{2g}+z\right)$ is called the total head.

5.4 VELOCITY HEAD

In equation (5.3), v is the velocity and $\left(\frac{v^2}{2g}\right)$ is called the velocity head.

5.5 PRESSURE HEAD

In equation (5.3), p is the pressure and $\left(\frac{p}{\rho g}\right)$ is called the pressure head.

5.6 DATUM OR POTENTIAL HEAD

In equation (5.3), z is called the datum head or potential head.

Note : All the head is nothing but energy per unit weight.

5.7 ASSUMPTIONS MADE FOR DERIVING BERNOULLI'S EQUATION

Following are some Assumptions made for deriving Bernoulli's equation :

(*i*) The flow is incompressible.

(*ii*) The flow is steady.

(*iii*) The fluid is non viscous ($\mu = 0$)

(*iv*) The flow is irrotational.

5.8 APPLICATIONS OF BERNOULLI'S EQUATION

Bernoulli's equation is widely used for the measurement of flow rate with the help of following flow rate measuring devices.

5.8.1 Venturimeter

Venturimeter is used to measure flow rate Q in m^3/s.

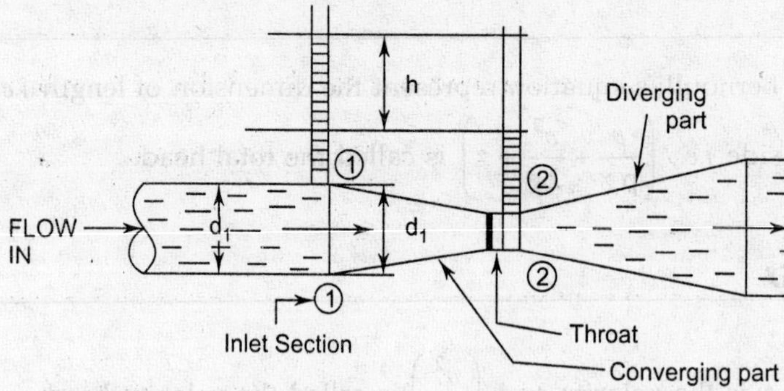

Fig. 5.2

Fig. 5.2 shows a pipe fitted with venturimeter which has three parts (*i*) converging part (*ii*) throat and (*iii*) diverging part.

Applying Bernoulli's equation in sections (1) and (2) we get

$$\frac{P_1}{\rho g} + \frac{v_1^2}{2g} + z_1 = \frac{P_2}{\rho g} + \frac{v_2^2}{2g} + z_2$$

Let us assume the venturimeter is horizontal.

So $$z_1 = z_2.$$

$$\frac{P_1}{\rho g} + \frac{v_1^2}{2g} = \frac{P_2}{\rho g} + \frac{v_2^2}{2g}$$

or $$\frac{P_1 - P_2}{\rho g} = \frac{v_2^2 - v_1^2}{2g}$$

or $$h = \frac{v_2^2 - v_1^2}{2g} \qquad ...(5.4)$$

From continuity equation

$$a_1 v_1 = a_2 v_2$$

or $$v_2 = \frac{a_1}{a_2} v_1.$$

From (5.4) $\Rightarrow$

$$h = \frac{\left(\frac{a_1^2}{a_2^2} - 1\right) v_1^2}{2g} = \left(\frac{a_1^2 - a_2^2}{a_2^2}\right) \frac{v_1^2}{2g}$$

or $$v_1^2 = \frac{a_2^2}{a_1^2 - a_2^2} (2gh)$$

Discharge $$Q_{Th} = a_1 v_1$$

or $$Q_{Th} = \frac{a_1 a_2}{\sqrt{a_1^2 - a_2^2}} \sqrt{2gh} \qquad ...(5.5)$$

Equation (5.5) is the expression for theoretical discharge. In actual discharge, there are some losses, $Q_{act} = C_d \times Q_{th}$.

or
$$Q_{act} = C_d \times \frac{a_1 a_2}{\sqrt{a_1^2 - a_2^2}} \sqrt{2gh} \quad ...(5.6)$$

where C_d is called the coefficient of discharge for venturimeter. $C_d < 1$.

If we use differential manometer

then
$$h = x\left[\frac{S_h}{S_o} - 1\right]$$

For
x = difference of heavier liquid in u-tube
S_h = *sp.* gravity of the heavier liquid
S_o = *sp.* gravity of the liquid flowing through pipe, here $S_o < S_h$.

If
$$S_o > S_l \text{ then } h = x\left[1 - \frac{S_l}{S_o}\right]$$

S_l = *sp.* gravity of lighter liquid.
x = difference of lighter liquid in u-tube.

For inclined venturimeter

$$Z_1 \neq Z_2$$

then
$$h = \left(\frac{P_1}{\rho g} + z_1\right) - \left(\frac{P_2}{\rho g} + z_2\right) = x\left[\frac{S_h}{S_o} - 1\right] \text{ or } \left[1 - \frac{S_l}{S_o}\right]x.$$

5.8.2 Pitot-tube

Pitot tube is used for measuring flow rate from velocity in a flowing fluid. It is L shaped. One end is open to atmosphere where as other end is kept immersed at about mid depth point where velocity is to be measured.

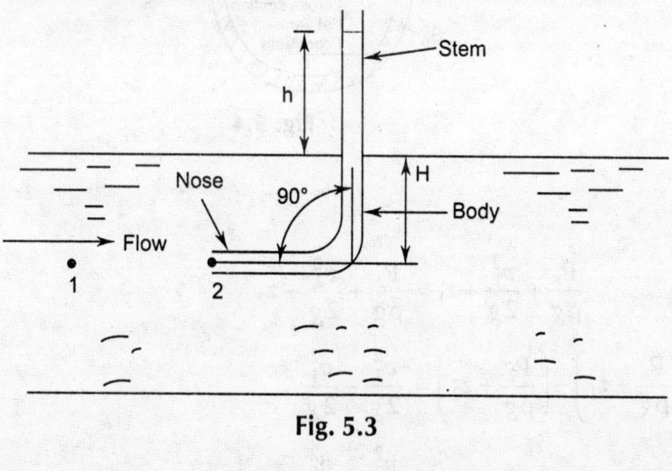

Fig. 5.3

Let us consider two points (1) & (2) in the fluid flow as shown in the figure. Applying Bernoulli's equation at sections (1) & (2) we get

$$\frac{P_1}{\rho g} + \frac{v_1^2}{2g} + z_1 = \frac{P_2}{\rho g} + \frac{v_2^2}{2g} + z_2$$

Now, $Z_1 = Z_2$, $v_2 = 0$ (at nose the kinetic head is converted to pressure head)

or, $$\frac{P_1}{\rho g} + \frac{v_1^2}{2g} = \frac{P_2}{\rho g}$$

or $$H + \frac{v_1^2}{2g} = H + h$$

or $$v_1 = \sqrt{2gh}$$

$$v_{act} = C_v v_1 = C_v \sqrt{2gh}$$ where

C_v = coefficient of velocity for pitot tube.

If a = cross sectional area of pitot tube, then Q_{act} is given by

$$Q_{act} = C_v a \sqrt{2gh} \qquad ...(5.7)$$

5.8.3 Orificemeter or Orifice Plate

Orificemeter is used to measure the rate of flow. Generally circular sharp edge plate is used in orificemeter.

Applying Bernoulli's equation in sections (1) & (2), we get,

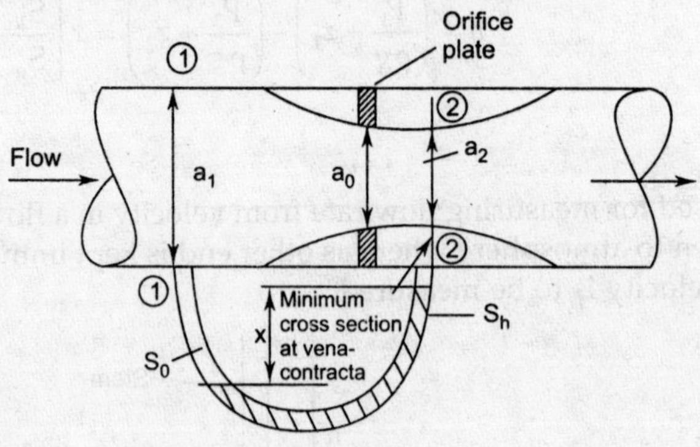

Fig. 5.4

$$h = x\left(\frac{S_h}{S_o} - 1\right)$$

$$\frac{P_1}{\rho g} + \frac{v_1^2}{2g} + z_1 = \frac{P_2}{\rho g} + \frac{v_2^2}{2g} + z_2$$

or $$\left(\frac{P_1}{\rho g} + z_1\right) - \left(\frac{P_2}{\rho g} + z_2\right) = \frac{v_2^2}{2g} - \frac{v_1^2}{2g}$$

$$h = \frac{v_2^2}{2g} - \frac{v_1^2}{2g} \qquad ...(5.8)$$

Now $$a_2 = C_c a_0 \text{ where } C_c = \text{coefficience of contraction}$$

$$a_2 v_2 = a_1 v_1 \text{ or } v_2 = \frac{a_1}{a_2} v_1 = \frac{a_1}{C_c a_o} v_1$$

From (5.8) $\Rightarrow$ $$h = \left(\frac{a_1^2}{c_c^2 a_0^2} - 1\right)\frac{v_1^2}{2g} = \left(\frac{a_1^2 - c_c^2 a_0^2}{c_c^2 a_0^2}\right)\frac{v_1^2}{2g}$$

or $$v_1^2 = \frac{c_c^2 a_0^2}{a_1^2 - c_c^2 a_0^2} 2gh$$

or $$v_1 = \frac{c_c a_o}{\sqrt{a_1^2 - c_c^2 a_0^2}} \sqrt{2gh}$$

$$Q = a_1 v_1 = \frac{c_c a_1 a_0}{\sqrt{a_1^2 - c_c^2 a_0^2}} \sqrt{2gh}$$

Let us simplify by using the following relation between C_c & C_d.

$$\sqrt{a_1^2 - c_c^2 a_0^2} = \frac{c_c}{c_d}\sqrt{a_1^2 - a_0^2}$$

Thus $$Q = \frac{c_d a_0 a_1 \sqrt{2gh}}{\sqrt{a_1^2 - a_0^2}} \quad ...(5.9)$$

where C_d = coefficient of discharge for orificemeter

$$(C_d)_{orifice} << (C_d)_{venturimeter}.$$

SOLVED EXAMPLES

Example 5.1. *A horizontal venturimeter with inlet diameter 30 cm and throat diameter 15 cm is used to measure the flow rate (discharge) of water. The pressure at inlet is 20×10^4 N/m^2 and the vacuum pressure at the throat is 25 cm of mercury.*

Find the discharge of water through venturimeter, take $C_d = 0.97$.

Solution.

Let inlet section is (1)

& throat section is (2)

$$\frac{P_1}{\rho g} + \frac{v_1^2}{2g} + z_1 = \frac{P_2}{\rho g} + \frac{v_2^2}{2g} + z_2$$

or $$\frac{20 \times 10^4}{10^3 \times 9.81} + \frac{v_1^2}{2g} = -0.25 \times 13.6 + \frac{v_2^2}{2g} \quad ...(1)$$

Now $$Q_{Th} = a_1 v_1 = a_2 v_2$$

or $$v_1 = \frac{Q_{Th}}{a_1} = \frac{Q_{Th}}{\frac{\pi}{4}(0.3)^2} = \left(\frac{14.15\, Q_{Th}}{1}\right)$$

or $$v_2 = \frac{Q_{Th}}{a_2} = \frac{Q_{Th}}{\frac{\pi}{4}(0.15)^2} = 56.59\, Q_{Th}$$

From equation (1) we get

$$\frac{20 \times 10^4}{10^3 \times 9.81} + \frac{(14.15\, Q_{Th})^2}{2 \times 9.81} = -0.25 \times 13.6 + \frac{(56.59\, Q_{Th})^2}{2 \times 9.81}$$

or $$Q_{Th} = 0.38\, m^3/\text{sec.}$$

$$Q_{act} = C_d\, Q_{Th} = 0.97 \times 0.38\, m^3/s$$

$$= 0.368\, m^3/s. \quad \textbf{Ans.}$$

Example 5.2. *Find the discharge of water flowing through a pipe 40 cm diameter placed in inclined position fitted with venturimeter having a throat diameter of 20 cm. The difference of pressure between the main and the throat is measured by a liquid of sp. gr 0.7 in an inverted u-tube which gives a reading of 25 cm. The loss of head between the main and throat is 0.25 times the kinetic head of the pipe.*

Solution.

$$\left(\frac{P_1}{\rho g} + z_1\right) - \left(\frac{P_2}{\rho g} + z_2\right)$$

$$= h = x\left(1 - \frac{S_l}{S_o}\right)$$

$$= 0.25\left(1 - \frac{0.7}{1.0}\right)$$

$$= 0.075\, m \text{ of water.}$$

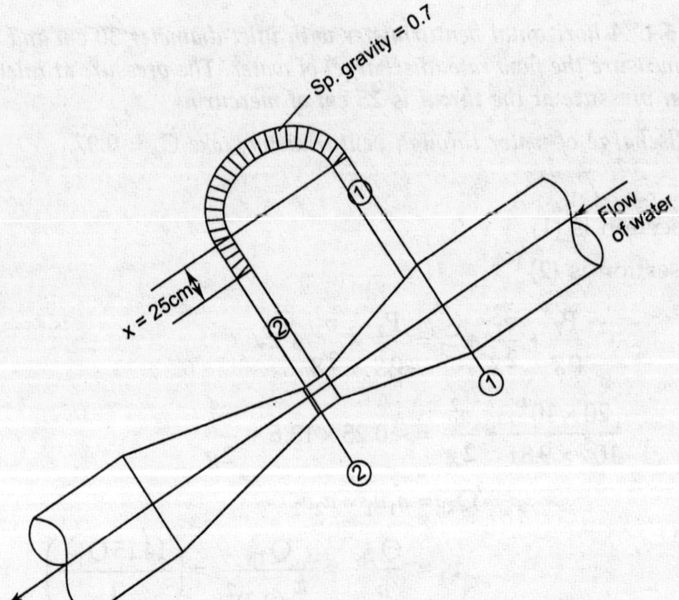

Fig. 5.5

Applying Bernoulli's equation in section (1) & (2)

$$\frac{P_1}{\rho g} + \frac{v_1^2}{2g} + z_1 = \frac{P_2}{\rho g} + \frac{v_2^2}{2g} + z_2 + \text{head loss}$$

or $$\left(\frac{P_1}{\rho g} + z_1\right) - \left(\frac{P_2}{\rho g} + z_2\right) = \frac{v_2^2}{2g} - \frac{v_1^2}{2g} + 0.25 \times \frac{v_1^2}{2g}$$

or $$0.075 = \frac{v_2^2}{2g} - 0.75\frac{v_1^2}{2g} \quad ...(1)$$

Now $$a_1 v_1 = a_2 v_2, \text{ or } v_2 = \frac{a_1}{a_2} v_1 = \left(\frac{0.4}{0.2}\right)^2 v_1 = 4 v_1$$

(1) ⇒ or $$0.075 = \frac{16 v_1^2}{2g} - 0.75\frac{v_1^2}{2g} = 15.25 \times \frac{v_1^2}{2g}$$

or $$v_1 = \sqrt{\frac{0.075 \times 2 \times 9.81}{15.25}} \; m/s$$

$$= 0.31 \, m/s.$$

$$\text{Discharge} = a_1 v_1 = \frac{\pi}{4}(0.4)^2 \times 0.31 = 0.039 \, m^3/s.$$ **Ans.**

Example 5.3. *A pitot tube is inserted in a pipe of 400 mm diameter. The static pressure in pipe is 150 mm of mercury (vacuum). The stagnation pressure at the centre of the pipe, recorded by the pitot tube is 1×10^4 N/m². Calculate the discharge of water through pipe, if the mean velocity of flow is 0.8 times the central velocity. (Take $C_v = 0.99$)*

Solution.

Diameter of pipe $= d = 400 \text{ mm} = 0.4 \text{ m}$

Area $a = \frac{\pi}{4}(0.4)^2$

Static pressure head $= 150$ mm of Hg (vacuum)

$= -0.15 \times 13.6$ of water

$= -2.04$ of water

Stagnation pressure $= 10^4 \, N/m^2$

$$= \frac{10^4}{1000 \times 9.81} = 1.02 \, m \text{ of water}$$

h = stagnation pressure head – static pressure head

$= 1.02 - (-2.04) = 3.06 \; m$ of water

Velocity at centre $= C_v \sqrt{2gh}$

$= 0.99\sqrt{2 \times 9.81 \times 3.06} \; m/s$

$= 7.67 \, m/s.$

Mean velocity $\overline{V} = 0.8 \times 7.67 = 6.136 \, m/s$

Discharge $Q = \bar{V} \times \text{area of pipe}$

$$= 6.136 \times \frac{\pi}{4} (0.4)^2 \; m^3/s$$

$$= 0.771 \, m^3/s. \quad \textbf{Ans.}$$

Example 5.4. *A pitot tube is properly placed just infront of a horizontally moving sub-marine. The axis of the sub-marine is 20 m below the surface of water. A u-tube differential manometer is connected along the axis of the sub-marine. The difference of mercury level is found to be 200 mm. Calculate the speed of the sub-marine,*

given $S_{sea\ water} = 1.026, S_{Hg} = 13.6.$

Solution. $$h = x\left(\frac{S_h}{S_o} - 1\right) = 0.2\left[\frac{13.6}{1.026} - 1\right] = 2.45 \text{ m of water}$$

velocity $$V = \sqrt{2gh} = \sqrt{2 \times 9.81 \times 2.45} = 6.93 \, m/s$$

$$= \frac{6.93 \times 60 \times 60}{1000} \text{ Km/hr} = 24.95 \text{ Km/hr.} \quad \textbf{Ans.}$$

Example 5.5. *The diameter of an orifice is 20 cm and it is used for measuring flow rate in a pipe of 40 cm diameter. A differential u-tube manometer reads 60 cm of mercury when it is placed on the two sides of the orificemeter. Calculate the discharge of oil (s = 0.89) through orifice.*

Take $C_d = 0.63$.

Solution. $$Q = \frac{C_d \, a_0 \, a_1}{\sqrt{a_1^2 - a_0^2}} \sqrt{2gh}$$

Now $$h = x\left[\frac{S_h}{S_o} - 1\right] = 0.6\left[\frac{13.6}{0.89} - 1\right] = 8.57 \text{ m of water}$$

$$a_0 = \frac{\pi}{4}(0.2)^2$$

$$a_1 = \frac{\pi}{4}(0.4)^2$$

$$Q = \frac{0.63 \times \frac{\pi}{4}(0.2)^2 \times \frac{\pi}{4}(0.4)^2}{\sqrt{\left\{\frac{\pi}{4}(0.4)^2\right\}^2 - \left\{\frac{\pi}{4}(0.2)^2\right\}^2}} \sqrt{2 \times 9.81 \times 8.57}$$

$$= \frac{0.0322}{0.1216} = 0.265 \, m^3/s. \quad \textbf{Ans.}$$

SOME OBJECTIVE TYPE QUESTIONS WITH ANSWERS

Choose the correct answers for the followings.

1. An open system is one in which

(a) heat and work cross the boundary of the system but the mass of the working substance does not.

(b) mass of working substance crosses the boundary of the system but the heat and work do not.

(c) both the heat and work as well as mass of the working substances cross the boundary of the system.

(d) neither the heat and work nor the mass of the working substances crosses the boundary of the system.

2. Which of the following is an intensive property of a thermodynamic system?
   (a) volume
   (b) temperature
   (c) mass
   (d) energy.
3. During throttling process
   (a) internal energy does not change
   (b) pressure does not change
   (c) entropy does not change
   (d) enthalpy does not change
   (e) volume change is negligible.
4. Kelvin-Planck's law deals with
   (a) conservation of energy
   (b) conservation of heat
   (c) conservation of mass
   (d) conversion of heat into work
   (e) conversion of work into heat.
5. If a heat engine attains 100% thermal efficiency it violates.
   (a) zeroth law of thermodynamics
   (b) first law of thermodynamics
   (c) second law of thermodynamics
   (d) law of conservation of energy.
6. For an irreversible process entropy change is
   (a) greater than $\delta Q/T$
   (b) less than $\delta Q/T$
   (c) equal to $\delta Q/T$
   (d) equal to zero.
7. The continuity equal at two sections 1 and 2 for an incompressible fluid is given as
   (a) $\rho_1 A_1 V_1^2 = \rho_2 A_2 V_2^2$
   (b) $\rho_1 A_1 V_1 = \rho_2 A_2 V_2$
   (c) $A_1 V_1 = A_2 V_2$
   (d) $\rho_1^2 A_1 V_1 = \rho_2^2 A_2 V_2$.
8. Euler's equation is written as

   (a) $\dfrac{dp}{\rho} + v^2\, dv + g\, dz = 0$

   (b) $\dfrac{dp}{\rho} + v\, dv + g\, dz = 0$

(c) $\frac{dp}{\rho} + v^2\, dv + g^2 dz = 0$

(d) $\frac{dp}{\rho^2} + v^2 dv + g\, dz = 0$

9. Kinematics viscosity is defined as equal

(a) dynamic viscosity × density

(b) dynamic viscosity/density

(c) dynamic viscosity/pressure

(d) pressure × density.

10. For pipes, turbulent flow occurs when Reynolds number is

(a) less than 2000

(b) between 2000 and 4000

(c) more than 4000

(d) less than 4000.

11. The change of entropy when heat is absorbed by the gas is

(a) positive

(b) negative

(c) positive or negative

(d) none of these.

12. Spherical shape of droplets of mercury is due to

(a) high density

(b) high surface tension

(c) high adhesion

(d) low vapour pressure.

13. Extensive property of a system is one whose value

(a) depends on the mass of the system, like volume

(b) does not depend on the mass of the system, like pressure

(c) is dependent on the path followed and on the state

(d) is always constant.

14. First law of thermodynamics furnishes the relationship between

(a) heat, work and properties of the system

(b) various properties of the system

(c) various thermodynamic processes

(d) heat and internal energy.

15. The more effective way of increasing the efficiency of a Carnot engine is to

(a) increase the upper temperature

(b) decrease the upper temperature

(c) increase the lower temperature

(d) decrease the lower temperature.

16. An increase in temperature results in

(a) increase in viscosity of gas

(b) increase in viscosity of liquid

(c) decrease in viscosity of gas

(d) decrease in viscosity of liquid

(e) both (a) and (d) above.

17. In an isothermal atmosphere, the pressure
    (*a*) remains constant
    (*b*) decreases linearly with elevation
    (*c*) increases exponentially with elevation
    (*d*) varies in the same way as the density.
18. A venturimeter is a device used to measure
    (*a*) pressure in a fluid
    (*b*) velocity at a point
    (*c*) flow rate
    (*d*) temperature of the fluid.
19. Process of heat transfer through a finite temperature difference is reversible.
    (*a*) True
    (*b*) False.

ANSWERS

1. (*c*)	2. (*b*)	3. (*d*)	4. (*d*)	5. (*c*)
6. (*a*)	7. (*c*)	8. (*b*)	9. (*b*)	10. (*c*)
11. (*a*)	12 (*b*)	13. (*a*)	14. (*a*)	15. (*a*) [because T_2 cannot be decreased]
16. (*e*)	17. (*d*)	18. (*c*)	19. (*b*)	

INDEX

Q

R

S

T

U

V

Z